*Genetics and the
Search for Modern
Human Origins*

Genetics and the Search for Modern Human Origins

John H. Relethford
State University of New York College at Oneonta

WILEY-LISS

A JOHN WILEY & SONS, INC., PUBLICATION

New York • Chichester • Weinheim • Brisbane • Singapore • Toronto

For ordering and customer service, call 1-800-CALL-WILEY.

Library of Congress Cataloging-in-Publication Data:

Relethford, John.
    Genetics and the search for modern human origins / John H. Relethford.
        p.   cm.
    Includes bibliographical references and index.
    ISBN 0-471-38413-5 (cloth : alk. paper)
        1. Evolutionary genetics.   2. Human population genetics.   I. Title.
    QH390 .R45 2001
    572.8′38—dc21

                                                                00-068633

Printed in the United States of America.
10 9 8 7 6 5 4 3 2 1

*To my wife, Hollie, and
my sons, David, Ben, and Zane,
for all their love,
and for putting up with my strange career.*

Contents

Chapter **1**

Reflections of the Past

"Where do you dig?"

I get this question a *lot*. I am a professor of biological anthropology,[1] the study of human biological diversity and evolution. Biological anthropology has several concerns. One is the study of biological differences and similarities among living human populations. How are human groups the same biologically? How are they different? Why are they similar or different? Biological anthropologists look at variation within and between living groups for clues to what happened in the past—the evolutionary history of our species. Some biological anthropologists focus on this question in a larger perspective by looking at similarities and differences between our species and other primates. Central to the study of comparative primatology is the question of the nature of humanity. How different are we from our closest relatives? This question concerns genetic, anatomic, and behavioral comparisons. Other biological anthropologists focus directly on remains from the past, primarily the fossil and archaeological remains of our ancestors. The study of the remains of our ancestors (known as paleoanthropology) deals with questions of origins and evolution. When and where did certain human characteristics, such as bipedalism and cranial expansion, occur? Why did these changes take place? How many different species lived at any given time, and which one is our ancestor?

This brief description is more or less the answer I give to the question "What is biological anthropology?" A follow-up question often concerns the nature of my research ("What are you working on?"). I explain that I have been involved in a wide range of research topics throughout my professional career, ranging from studies of migration in colonial America to studies of the genetic history of Ireland. During the past decade, my main interest has been in the recent evolution of modern humans, specifically the past 200,000 years or so. I am interested in where our recent ancestors came from. We know that early humans first left Africa almost two million years ago, moving into parts of Asia and Europe. There is a continuing controversy about our relationship to these early humans. Were all of these groups among our direct genetic ancestors, or only some of them, or perhaps only one of them? Did these early humans consist of one species, two species, or more? Were the Neandertals of Europe and the Middle East among our ancestors, or were they a side branch in human evolution? When and where did our modern anatomy appear? Was it in one place at one time, or something that resulted over long periods? These are just some of the questions posed by those of us interested in recent human evolution.

Any interest in the study of human evolution normally (and naturally) conjures up images of dusty and dirty anthropologists sweating under the sun while unearthing the bits and pieces of our ancestors that we use to reconstruct our species' history. Other images might include the archaeologist digging up ancient stone tools or crawling through caves looking for ancient paintings. Mention an interest in human prehistory or evolution and we all tend to think of scientists digging in remote lands, including both fictional individuals (Indiana Jones, for example) and those in real life (the entire Leakey family, or Donald Johanson of "Lucy" fame). Given these images, it is reasonable to ask any anthropologist doing research on human evolution the logical question, "Where do you dig?"

My answer is usually a disappointment—I don't dig. Actually, most anthropologists involved with human evolution are not directly involved in fieldwork but work on the overall puzzle in other ways. The study of human evolution is truly interdisciplinary and involves professionals in the fields of anthropology, paleontology, geology, animal behavior, cultural anthropology, and others. The questions we ask are often broad in nature and require input from a variety of fields. It is no longer possible or even desirable for a single person, no matter how well trained, to do it all on his or her own. The contributions of many fields are needed to put the pieces together and answer even the most basic questions.

The study of human evolution is a multidisciplinary endeavor involving different stages of collection, analysis, and interpretation. The raw data frequently consist of fossil and archaeological specimens that are discovered and excavated. To do this, the expertise of many disciplines is required including anthropology, vertebrate paleontology, geology, and paleobotany, among others. The skills needed for initial excavation are quite varied, including everything from archaeological methods to aerial photography.

The multidisciplinary approach does not end with the initial fieldwork. Specialists in anatomy are needed to clean, study, and describe the fossils. Anatomic and statistical analysis will be used to classify the fossils and compare them with previous discoveries. Archaeologists will then pore over stone tools or any other evidence of early human behavior such as fire and hunting. Specialists in dating methods will work with geologic samples to assign a date to the site. Information on the past environment will come from experts in analysis of the remains of other animal and plant remains, ranging from fossil pollen to pig's teeth. The initial fieldwork is vital to any understanding of human evolution, but it is only part of what goes on. Other researchers busy themselves in laboratories or in front of a computer (which, incidentally, is where I spend most of my time).

Whether one works in initial fieldwork or subsequent analysis back in the labs, there have been two primary sources of information used to answer questions of human origins and evolution—the fossil record and the archaeological record (or, as referred to by some anthropologists, "bones and stones"). So when I tell people that my research deals with human evolution, the logical conclusion is that I am involved primarily with either bones or stones and that I either dig them up or analyze them in the lab. Actually, I work with a third source of information on human evolution—the genetics of living human populations. My main tool is not a shovel or a microscope, but a computer. I use genetic data obtained from living human populations to test hypotheses and make inferences about the history of these groups. The basic research strategy is to examine patterns of genetic variation in living human populations and make inferences about the evolutionary patterns that gave rise to the observed patterns. The present is the key to the past, or in terms of the title of this chapter, genetic variation today is a *reflection of the past*.

1.1   THE GENETIC HISTORY OF THE HUMAN SPECIES

Many studies of genetics and population history have been carried out by anthropologists and geneticists in populations across the world. Some studies have focused on cultural isolates within larger societies. Others have looked at the history of larger regions, including studies of nations and continents. Until the 1980s, far fewer studies attempted to reconstruct the long-term history of the entire human species. The initial focus on local populations led anthropologist Henry Harpending to write in 1974 that such studies "have not advanced our understanding of human evolution in a global sense."[2] Although some studies did look at global history, Harpending was correct in noting that the emphasis had been primarily on small local populations.

This situation changed most dramatically in the late 1980s. A key event was the publication in 1987 of a short article in the journal *Nature* entitled "Mitochondrial DNA and human evolution." Written by Rebecca Cann, Mark Stoneking, and Allan Wilson, this article compared DNA sequences from human mitochondria collected on 147 individuals with ancestors from Europe, Africa, East Asia, Australia, and New Guinea.[3] One of their most controversial findings was the reconstruction of a genealogy of DNA types that suggested that all living humans had a common female ancestor that lived in Africa between 140,000 and 290,000 years ago. Unlike nuclear DNA (the DNA that is contained in the 23 chromosome pairs inside the nucleus of our cells), mitochondrial DNA is inherited only through the mother's line. We all have our mother's mitochondrial DNA, she has her mother's mitochondrial DNA, and so on. If we assume that the only changes over time that can happen result from mutation, and if we further can make a reasonable estimate of the rate at which these mutations occur, then such DNA genealogies can be used to estimate where and when our common female ancestors lived. In this case, the evidence suggested we all have a female ancestor in common and that she lived in sub-Saharan Africa roughly 200,000 years ago. Given the image of a single female ancestor, it was only a matter of time before the media and the scientific community began referring to this ancestor as "Eve."

The evidence from this, and other, studies of mitochondrial DNA, and the alternative interpretations, are discussed in more detail in later chapters. For the moment, I only want to point out that this analysis is part of an ever-increasing number of studies that have relied on genetic data for inferences about our species' history—explaining the past by looking at the present. We can study patterns of human genetic variation as they exist in the world today, noting which groups are most similar to which others, and for which characteristics. The underlying question is "Why?" Are certain groups more similar genetically because they have recently shared genes or because they live in similar environments? When we can show that groups are closely related, the next question is "How closely related?" Are we seeing evidence of relationships that occurred fairly recently, perhaps within the past few thousand years, or more distantly into the past, perhaps on the order of hundreds of thousands of years? To answer these questions and others, we must use genetic information on living human populations as only one of several clues as to the distant history of our entire species. The genetic evidence must also be considered alongside other clues to our past, primarily the fossil and archaeological evidence.

A more complete description of the debate over modern human origins is given in later chapters. For the moment, as a way of introducing the overall structure of this book, I provide a summary of the basic debate. Today, there is ample evidence from the fossil and archaeological records for human populations in Africa, Asia, and Europe over the past million years or so. Humans, and human ancestors, first evolved in Africa. By two million years ago, there is evidence of the origin of human ancestors in Africa that had relatively large brains (though not initially as large as ours on average). These ancestors also had human body proportions and were tool users. They began including animal protein in their diet, first by scavenging and then later with hunting. Anthropologists generally refer to these early humans by the species name *Homo erectus*, although not everyone agrees on the name. Some populations of *H. erectus* moved out of Africa and moved east, eventually reaching modern-day Indonesia. Recent geologic dating suggests that *H. erectus* arrived in Southeast Asia perhaps as early as 1.8 million years ago and into the eastern fringes of Europe by 1.7 million years ago.

In any event, by half a million years ago, human ancestors were living in parts of Africa, Asia, and Europe. These populations were in some ways different from the older specimens labeled *Homo erectus*, including a further increase in brain size. By several hundred thousand years ago, the average brain size was similar to that of modern humans, although the skulls were still somewhat differently shaped, having large brow ridges, larger faces, and a somewhat lower skull relative to living humans. These fossils are clearly "human" in the broad sense in terms of both biology and behavior, but it is also clear that they are different in some ways compared to living humans. There is considerable debate over what to call these specimens. Some argue that they are evidence of an earlier evolutionary stage of ourselves, similar yet different in some ways. As such, they are our direct ancestors and perhaps should be labeled as belonging to our species—*Homo sapiens*. Some scholars argue that it is useful to attach a descriptive term that simultaneously acknowledges their similarity to and difference from us. One such term that has been used is "archaic *Homo sapiens*." Other anthropologists suggest that not all of these early humans were our ancestors. According to this view, these specimens represent two or more different species, of which only one is our ancestor and the others are now extinct. Some of the different names suggested here are *Homo heidelbergensis* and *Homo neanderthalensis*. Still other anthropologists argue that because we are not yet sure about whether these fossils belong to one or more species we should simply refer to them as "archaic humans," an approach I use throughout this book.

This is not just a matter of arguments about names. The designation of different forms by different species names reflects one's views on their evolutionary relationship. The most important question, forming the main focus of this book, is the relationship of these archaic humans with specimens that are more clearly modern in anatomy. Quite simply, we see evidence of *Homo erectus* going back two million years, with later fossils found in Asia and Europe. By several thousand years ago, we see evidence of archaic humans throughout the Old World (Africa, Asia, Europe). By roughly 30,000 years ago, all fossils found in the Old World can be typically described as "anatomically modern." Apart from the complex arguments over names, what does this fossil record tell us? At the simplest level, we see a clear change over time, primarily reflected in the evolution of larger brains and smaller faces and teeth, along with increasing complexity of stone tool technology and culture.

The arguments come when we look more closely at this picture, trying to see exactly what happened in different places and at different times. What is the relationship between populations of *Homo erectus*, archaic humans, and modern humans in different geographic regions over time? How many species existed? Did more than one species exist at a given point in time? If so, then which species is our ancestor, and where and when did they live? These questions form the crux of the modern human origins debate.

A number of models have been proposed to answer these questions. The specifics of these models are discussed in Chapter 3, but for the moment I wish to briefly identify the two major classes of models—replacement and multiregional evolution. The replacement model postulates that modern humans first arose as a new species (*Homo sapiens*) roughly 150,000 to 200,000 years ago in Africa. This new species began to expand in its geographic range, and by 100,000 years ago some populations had begun to spread out from Africa into other regions, arriving first in the Middle East and then later moving into Australia, East Asia, and finally Europe. According to this model, the archaic humans that were already living outside of Africa were replaced and became extinct. The cause of this replacement is not clear but is often presumed to be some biological or technological advantage that allowed this new species to replace earlier humans.

The multiregional model takes a different view. Here, all humans over the past two million years belonged to a single evolutionary line. Human populations began expanding out of Africa close to two million years ago. Regional differences formed over time, but never enough to cause a "split" leading to a separate species. Humans in different geographic regions remained connected through occasional migration that promoted the sharing of genes and culture. Over time, the human species changed both biologically and culturally, and our use of labels such as *Homo erectus*, archaic humans, and modern humans are just that—labels denoting different stages in the evolution of our species.

To contrast these two models, it is useful for you to consider the question of where your ancestors lived at different points in time. Your most immediate ancestors are your parents, followed by your grandparents, followed by your great-grandparents and so forth. Where did they live? This question can often be answered for at least one or two generations; any earlier and we rarely have genealogical information on exactly *who* all of our ancestors were, let alone where they lived. For example, I have information on one of my ancestors who was born in England in 1595, but this is only one of the thousands of my ancestors that lived that long ago. I simply do not know who the rest were, or where they lived, because of the necessary incompleteness of most family records. As we move further back into the past, reconstruction of genealogies becomes even more difficult, and once we go back into the past beyond written records it becomes impossible to identify specific ancestors.

We can try, however, to look at general patterns of historical relationship between populations and make tentative suggestions about past ancestry. As a simple thought experiment to illustrate the modern human origins debate, consider the following question—Where did your ancestors live 200,000 years ago? According to the replacement model, *all* of your ancestors this far back lived in Africa, regardless of whether your *recent* ancestry was from Europe, Asia, Australia, or Africa. If the replacement model is correct and if we had records this far back, we would find that each and every one of your many thousands of ancestors lived in Africa at the time.

How about the multiregional model? According to this view, your ancestors 200,000 years ago would be more scattered. *Some* might have lived in Africa, whereas others might have lived in Europe, Asia, or elsewhere. If the multiregional model is correct, our hypothetical records would show a great deal of geographic variety. The exact mix of ancestral locations would vary from one person to the next, but the general pattern would be to have ancestors from all over. The multiregional model further suggests that this pattern would be the same at earlier times, going back all the way to the first movement of *Homo erectus* out of Africa. At that point in time (two million years ago), all of your ancestors would have lived in Africa, but afterwards some would have lived in other regions.

Of course, we do not have these hypothetical written records. If we did, there would be no argument. Instead, we have to approach the question of ancestry by looking at other sources of information—the fossil record, the archaeological record, and the genetics of living people. Whatever happened in the past, it affected human populations and the genes they passed down generation to generation, arriving ultimately in the present. Although we lack written records that far back, we still retain the signature of past events in our genes. Our task then is to reconstruct our species' history by examining these reflections of the past.

1.2   THE ORGANIZATION OF THIS BOOK

This book is intended primarily for advanced undergraduate students, graduate students, and professionals but, I hope, can be appreciated by anyone interested in this fascinating field. No prior background in genetics or evolution is needed. The task of this book is to examine the modern human origins debate primarily from the evidence we now have on the genetics of living (and relatively recent) human populations. To do this, it is first necessary to provide some basic review of genetics and a more detailed review of the fossil record and the debate. Chapter 2 provides some basic background in genetics and evolutionary theory to discuss *how* we go about reconstructing history from genetic data. This chapter also provides two examples, one focusing on the history of differences *between* species and the other an example of history *within* a species. Chapter 3 returns to the modern human origins debate by reviewing the fossil record of our ancestors over the past few million years and examining the debate, both historically and currently, in greater detail.

Chapter 4 turns to the first of several lines of genetic evidence that is relevant to the search for modern human origins. This chapter looks at reconstructing genealogies from genetic data ("gene trees") and what such trees tell us (and what they don't tell us). Chapter 5 examines the question of population diversity, that is, how much genetic variation exists in different parts of the world. For most genetic traits, humans living today in sub-Saharan Africa show the highest amount of genetic diversity in the world. This tells us something about the evolutionary past of human populations, but what? Chapter 6 turns to variation *between* populations by examining the degree and pattern of genetic differences between people living today in different parts of the world. Chapter 7 looks at our history demographically by examining methods that estimate how many people lived in the past and where they lived. Chapter 8 looks at recent genetic evidence from actual fossils, made possible by the recent (1997–2000) extraction of DNA from ancient Neandertal fossils. Chapter 9 pulls all these pieces together, lays out the pros and cons of different modern human origin models, and provides my attempt at possible resolution.

In a discussion of any controversial topic it is necessary to explain one's own perspective ("Where am I coming from?"). Over the past few years, I have drifted back and forth in my own interpretation of which model, replacement or multiregional, is correct. Part of my shifting interpretation has had to do with my own research into this question, and part is due to the influence of the work of others. My interpretations, and those of others, are likely to change as new evidence, be it genetic, fossil, or archaeological, continues to accumulate. This is not unexpected—it is the very nature of science to be tentative. All one can do is evaluate how well the evidence fits different hypotheses at a given point in time. Even when the evidence clearly supports one interpretation over another, the scientist must remain open to change as new evidence enters the picture.

It is unfortunately all too common to read newspaper headlines that proclaim something to the effect that some new discovery has overturned previously held ideas. This happens in anthropology but also in astronomy, geology, and all sciences. To those unfamiliar with how science works, this must look devastating. After all, how can one place faith in science if the scientists admit that they were wrong? What fools they must be! While understandable, such a reaction fails to grasp the true nature of science, not a collection of facts and truths but a general method of learning the truth. Hypotheses are formed and then tested. If your hypothesis is tested and found to be incorrect, it must be modified or rejected. Human nature being what it is, this is often a painful process, and many are understandably reluctant to let go of an idea past its time. However, science is self-correcting and will tend to weed out our attempts to hold on to something past its time (or, if we don't do it, someone else will!). Yes, scientists make mistakes in the sense that hypotheses can be overturned. This is not abnormal; indeed, it is the normal nature of the scientific process.

At various times during the past decade I have been convinced that the replacement model of modern human origins was correct. At other times, I was convinced beyond a doubt that a multiregional interpretation was the correct one. At this time, I still lean more toward the view that our ancestry is mostly, but not exclusively, out of Africa—a variant of the multiregional interpretation (and I will explain why)—but it is less a firm conclusion than a statement of relative probability. There is still much more work to be done in the field, and I am not as sure as many of my colleagues that things are definitive one way or the other. Part of the problem is that new evidence, particularly genetic, continues to accumulate. It is perfectly reasonable to support one hypothesis over the other based on current evidence, but one must still be open to future developments that might change things.

The way in which science works seems at times completely foreign to our general inclination, particularly in American culture, to line up definitively behind one idea or the other. Consider, for example, election campaigns. You might hear one candidate argue that lowering taxes will lead to future economic growth, while another candidate might argue the reverse. You have to examine the evidence and evaluate the likelihood of both statements before making your decision as to who receives your vote. What you are less likely to hear is a candidate who makes the statement that, given current evidence, he or she supports one hypothesis but is willing to admit the possibility of being wrong and will continue to look at new evidence. Such a statement would likely be regarded as "wishy-washy" and "not standing up for one's beliefs." I suspect that this candidate would lose.

Scientific progress operates differently. When a scientist makes a statement such as, "The genetic data prove an African replacement model." it is not accepted automatically. It is the very nature of science to be critical and to look for alternative explanations. A major purpose of this book is to point out alternative interpretations of the genetic evidence for modern human origins. It is very common to read statements in the literature that suggest that a given set of genetic evidence is compatible with the replacement model. For many of these studies, I agree—the data *are* often (although not always) compatible with the replacement hypothesis. However, compatibility with a hypothesis is not *proof* of a model unless the data are incompatible with all alternatives. Compatibility only establishes the *possibility* that a given hypothesis is correct.

Consider a simple example. You observe me getting out of my car with a bag of groceries. You could logically infer that I am returning from the supermarket. However, suppose you formulate the hypothesis that I have come directly from the supermarket. The evidence (my bag of groceries) is certainly compatible with this hypothesis, but it does not prove it, because there are other possibilities. An alternative hypothesis might be that I stopped at a video rental store on my way home from the supermarket. The evidence is also compatible with this hypothesis. To determine which hypothesis (if either) is correct, you would need to look for a way to rule out one hypothesis in favor of the other. In this case, I can't think of any sure-fire way to do this short of having actual witnesses to my travels.

This book considers evidence, both pro and con, for both replacement and multiregional models. As I write this, I lean one way more than the other but am not completely convinced as to which interpretation is correct. My wanderings through the evidence throughout this book may not answer the question, but the main purpose is to show *how* we go about answering the question and to make some suggestions to break the log jam. I certainly hope that the definitive answer will come soon in our lifetimes, even though it means this book is likely to become dated quickly. On the other hand, the lessons we learn from the search for human origins may persist. I certainly hope so.

A note about the format of this book: To make the text as accessible as possible I have not included citations and other footnotes in the body of the chapters but at the end of the book under "Chapter Notes." These notes include both bibliographic information as well as additional discussion where needed. Mathematics is kept to a minimum in the main text, with further details given in the chapter notes or the cited literature.

Finally, there are many colleagues whom I wish to thank. This list includes all of those whom I can recall having one (or many) conversations regarding the modern human origins debate. Some of them agree with my views, while others agree with very little. Nonetheless, they have all been extremely helpful in my past research and writing and have had an influence on my thinking (even when it ended up with different conclusions). Needless to say, I alone am responsible for any errors or misrepresentations. Thanks go to Stan Ambrose, Guido Barbujani, John Blangero, C. Loring Brace, Rachel Caspari, Elise Eller, John Fleagle, Dave Frayer, Mike Hammer, John Hawks, Rosalind Harding, Henry Harpending, Jody Hey, Bill Howells, Lynn Jorde, Richard Klein, Lyle Konigsberg, Ken Korey, Andy Kramer, Phil Rightmire, Alan Rogers, Tad Schurr, Steve Sherry, Fred Smith, Anne Stone, Mark Stoneking, Chris Stringer, Alan Templeton, Alan Thorne, Sarah Tishkoff, Eric Trinkaus, Milford Wolpoff, Ken Weiss, and Catherine Willermet. If I have inadvertently left someone out, I offer my sincere apologies—chalk it up to a failing memory.

I also thank Kenneth A. Korey (Dartmouth University), Andrew Kramer (University of Tennessee), Michael Hammer (University of Arizona), Alan Rogers (University of Utah), Jody Hey (Rutgers University), and Kenneth K. Kidd (Yale University) for reviewing a draft of this book; I truly appreciate your time and assistance and your suggestions for improvement. I am also grateful to Luna Han, Assistant Editor of *Life Sciences*, for having suggested this project and for having given continual support and encouragement through its development. Thanks also to Danielle Lacourciere, Associate Managing Editor, for helping me through the production phase and putting up with my silly questions.

Chapter 2

Evolution and Genetic History

Popular culture often shapes our images of genetics and genetic analysis. We often think genetic questions are answered quickly by the appropriate laboratory methods. Find a strand of hair, analyze it with the right equipment, and push a button—out pops the name of the killer. Extract ancient DNA from some prehistoric amber and reconstruct a dinosaur. Analyze the mitochondrial DNA of an ancient human and determine whether they belonged to our species. Some examples are based on fact and others are fictional, but all fit with a common perception that the final product of a genetic analysis, such as a printout of the DNA sequences, provides us with a quick and easy answer. In reality, the process is much more complicated. The interpretation of the results of any scientific investigation often represents the hardest part of a study. The ultimate objective is to answer the question, "What do these results mean?"

To understand how genetic data can be used to answer questions concerning history, we must first understand what kinds of genetic data can be collected from living human populations. What is it exactly that we measure? What do these data tell us about the patterns of genetic variation that exist within and between living populations? The second thing we need to understand is how we explain these variations; specifically, what sequence of evolutionary events could have given rise to the observed patterns of genetic variation? In terms of the modern human origins debate, this means we must evaluate the likelihood of different replacement and multiregional models. To do this, we must further understand how evolution works, what the predictions of the models are, and how well they fit the observed data.

This chapter provides the basic background in evolutionary theory necessary to follow the major arguments over the genetic evidence for modern human origins.[1] First, I review some basic genetics and then move on to a brief description of the kinds of genetic data that are routinely collected and analyzed. I then move on to a review of basic evolutionary theory, focusing on the four mechanisms of evolutionary change: mutation, natural selection, genetic drift, and gene flow. This chapter concludes with two case studies involving the reconstruction of history from genetic data, one looking at genetic differences within a species (the population history of Ireland) and the other looking at differences between species (the split of ape and human lineages).

2.1   THE GENETIC CODE

We are all familiar with the term DNA and how it is referred to as "the genetic code." Hardly a day goes by without hearing about DNA in popular culture (e.g., the *Jurassic Park* movies) or in the news (as I wrote this, scientists were discussing the possibility of extracting DNA from a 20,000-year-old corpse of a prehistoric mammoth). The acronym DNA stands for **d**eoxyribo**n**ucleic **a**cid, a molecule that contains the instructions for biological structures. The DNA molecule consists of two strands that form a double helix and is made up of nucleotides, each of which consists of a sugar, a phosphate group, and a nitrogen-containing base. There are four bases: adenine (A), guanine (G), thymine (T), and cytosine (C). The two strands are complementary—the nucleotide base A always pairs with the base T, and the base G always pairs with the base C. Each position along the DNA molecule thus consists of a pair of bases, referred to as a base pair (and abbreviated as "bp"). The human genome (all of the DNA in the human species) is made up of approximately three billion base pairs.

The four bases (A, G, T, C) provide a four-letter "alphabet" for specifying the genetic code. Biological structures are primarily made up of proteins, which in turn are made up of amino acids. A sequence of three bases provides the code for an amino acid. For example, the DNA sequence CTT codes for the amino acid glutamic acid, and the sequence CGA codes for the amino acid alanine. There are 64 possible 3-letter combinations of the 4 different bases, which is more than enough to provide the genetic code for the 20 amino acids required for protein synthesis. Some code sequences provide "punctuation" for the genetic message (e.g., ATT codes for "stop," indicating the end of a message). Other code sequences are redundant, so that the same amino acid can be specified by slightly different DNA sequences. The amino acid valine, for example, can be specified by the sequences CAA, CAT, CAC, and CAG. On the other hand, some amino acids have only one associated DNA sequence, such as tyrosine, which is coded for by the sequence ATA.

It may seem remarkable at first that much of the diversity of life can be generated with such a simple code relying on four letters. However, consider the English language; only 26 letters are used, but they can be combined into different words, sentences, and paragraphs to produce everything from a term paper to a love letter. As an even more basic example, consider computers. Everything that we do on computers, from playing games to sending e-mail, ultimately is the product of a binary coding system that has only two numbers, "0" and "1." Different sequences of zeroes and ones code for different operations inside a computer chip. Machine languages consist of different arrangements of these basic operations, and machine languages can in turn be used to produce higher-level languages, which in turn are used to write the computer programs that we use.

All of life rests upon a fundamental property of the DNA molecule—it can make copies of itself. Because each strand of the DNA molecule consists of complementary bases, one strand can attract free-floating bases to make a copy. This is a complex biochemical process, but it essentially consists of the separation of the DNA molecule into two separate strands, each of which can attract free-floating bases to form two copies. Consider for example one strand with the sequence CCA. Because G pairs with C and T pairs with A, the complementary strand has the sequence GGT. Once separated, the strand with the sequence CGA will form the complementary strand GGT. The other strand will form the complementary strand CCA. The net result is two identical DNA molecules. The ability to make copies provides a means by which the genetic code is passed on from cell to cell.

DNA provides the code for protein synthesis with the aid of a related molecule known as RNA (ribonucleic acid) that serves as the messenger for the genetic instructions. This is a complex biochemical process involving a particular form of RNA known as messenger RNA traveling to the site of protein synthesis, where a form known as transfer RNA carries out the synthesis. The mechanics of this process are beyond the scope of this book but are described in most general genetics texts.

Your entire DNA collectively defines your genome. Most of your DNA is contained inside the nucleus of cells in long strands known as chromosomes (some DNA, to be discussed later in this chapter, is found within the mitochondria of cells). In body cells, chromosomes are found in pairs. Different species have differing numbers of chromosome pairs; for example, fruit flies have 4 pairs, chimpanzees have 24 pairs, and humans have 23 pairs. The DNA in the chromosomes contains not only the genetic instructions for the production of proteins but also a large amount of DNA that has no functional significance. The term "gene" is usually reserved for a DNA sequence that codes for an RNA product or a polypeptide chain—a molecule made up of many amino acids. For example, the hemoglobin molecule in your blood, which carries oxygen throughout your body, is coded for by several hemoglobin genes, which specify the protein structure of hemoglobin. One of the most interesting findings of recent genetics is the realization that much of our DNA sequences are noncoding. Even the coding genes contain portions carrying the actual code for production of polypeptides (exons) as well as portions that do not code for any protein (introns).

We refer to the actual position of a gene or other DNA sequence on a chromosome as a locus (the plural form of this word is "loci"). At each locus, you have DNA from both your mother and your father. Sometimes the genes or DNA sequences will be the same, but sometimes they will be different, because each species contains different variants of genes and DNA sequences. These variant forms are known as alleles.

2.2 THE PROCESS OF INHERITANCE

The DNA molecule has the ability of self-replication; it can make copies of itself. In this way, genetic information is transmitted from cell to cell, as well as from generation to generation. The process of replication is different in body cells and sex cells. In human body cells, the process of mitosis produces two identical cells, each

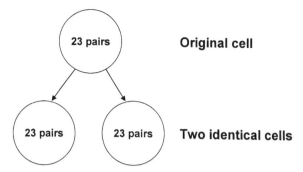

Figure 2.1. *Mitosis, the replication of body cells. Each cell contains 23 chromosome pairs. During mitosis, two identical cells are created, each with 23 chromosome pairs.*

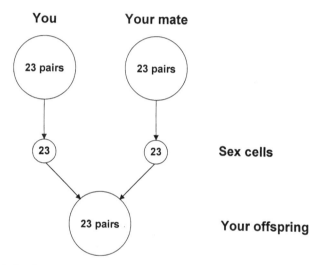

Figure 2.2. *Meiosis, the replication of sex cells. Each parent has 23 chromosome pairs. During meiosis, only one chromosome from each pair winds up in a sex cell. The same process occurs in the sex cells of your mate. When you and your mate conceive, the fertilized egg (zygote) now has the full complement of 23 chromosomes. Half of the offspring's DNA came from you, and the other half came from your mate.*

with 23 pairs of chromosomes (see Fig. 2.1). In this way, all of the genetic information is transmitted as body cells replicate. The situation is different for sex cells (sperm cells in the man, eggs in the woman). Here, only one of each chromosome pair is replicated in the sex cell, a process known as meiosis. After fertilization, the new offspring has the full complement of chromosome pairs (see Fig. 2.2). It is important to note that you only pass on *half* of your genetic code—one chromosome from each pair—and your offspring receives *half* of his/her genetic code from both you and your mate.

Before the discovery of DNA, scientists had already worked out many of the basic principles of inheritance. The first person to do so was Gregor Mendel, whose nineteenth century experiments on pea plants uncovered several basic principles. Before Mendel, a common misunderstanding was that genetic information in the sperm and egg somehow blended together. Mendel instead showed that discrete units (genes) were inherited and that the physical manifestation of these genes depended on which forms (alleles) were inherited and how they interacted. For example, one of Mendel's experiments examined seed color for the pea plants. He crossed plants with yellow seed color with plants with green seed color. Under the idea of blending inheritance, one might expect that the offspring would have a seed color that blended the colors of yellow and green. Mendel found that all of the offspring had yellow seed color, which was a clue that somehow the effects of one allele (yellow seed color) dominated in effect over the other allele (green seed color).

2.2.1 Dominant, Recessive, and Codominant Alleles

We now know that genetic information is inherited from both parents. At any given locus for nuclear DNA (the DNA in the nucleus of the cell), you have two sets of instructions, one from your mother and one from your father. The combination of these two determines your genetic makeup, known as your genotype. If you have the same allele from both parents the genotype is termed homozygous, whereas if the alleles are different the genotype is termed heterozygous. The actual observable appearance is known as the phenotype, which depends on whether one allele is dominant (showing its effect) or recessive (having its effect masked). In addition, some alleles can be codominant, in which case both alleles are expressed.

As an example of these concepts, consider the ability to taste the chemical phenylthiocarbamide (abbreviated PTC). Some people can taste PTC, whereas others cannot. One of the factors affecting a person's ability or inability to taste PTC is a gene with two alleles, usually labeled as "T" (for "taster") and "t" (for "nontaster"). The T allele codes for the ability to taste PTC, and the t allele has a different code resulting in the inability to taste PTC. Given two alleles in our species, everyone has one of three different genotypes: TT, Tt, or tt. People with the TT genotype have two copies of the allele that allows tasting; thus, their phenotype is that of a "taster." Likewise, people with the genotype tt have two copies of an allele that results in their inability to taste PTC; thus, their phenotype is "nontaster." The heterozygotes, those with the genotype Tt, have one copy of each allele. We have determined through observation that the T allele is dominant and the t is recessive, and people with the Tt genotype express the dominant allele and are "tasters."

This example illustrates more than just the process of dominance. It also shows that we can discuss genetic diversity at several different levels. The most noticeable level is that of the phenotype; a simple test results in classifying people into two groups—"tasters" and "nontasters." There is another level of diversity, that of the genotype. Although all "nontasters" have the tt genotype, the "tasters" could have either the TT or Tt genotype. Classification of people on the basis of phenotype would obscure the underlying genetic diversity. Recessive alleles can be hidden in one generation. A person with genotype Tt would phenotypically be a "taster" but could pass on the nontasting allele t to the next generation. If both parents had the genotype Tt they would both be "tasters" but have a one in two chance of each passing on the t allele, so that there is a one in four chance a given child could have the genotype tt and have a different phenotype than the parents.

In many cases, different alleles are codominant. A good example in humans is the MN blood group, named after the types of molecules (M or N) present on the surface of red blood cells. There are two common alleles: *M*, which codes for type M molecules, and *N*, which codes for type N molecules. There are three genotypes: *MM*, *MN*, and *NN*. People with the *MM* genotype obviously have phenotype M, and people with the *NN* genotype obviously have phenotype N. What about people with the heterozygous genotype *MN*? In this case, we know from observation that the *M* and *N* alleles are codominant, and *both* effects are shown. A person with the *MN* genotype would have phenotype MN, where both M and N molecules are present on the surface of his or her red blood cells.

2.2.2 Inheritance of Chromosomes

Alleles come in pairs, but only one is passed on to a given offspring. You pass on one of each chromosome pair through a random process. The probability of passing on a particular chromosome is 50%. This probability is independent for each chromosome pair, a principle known as Mendel's law of independent assortment. Imagine, for example, an organism with two pairs of chromosomes. Through the process of meiosis, only one of each pair is passed on in a given sex cell. If we label the two chromosomes of the first chromosome pair as 1-1 and 1-2, and the two chromosomes of the second chromosome pair as 2-1 and 2-2, we see that there are four possible combinations of sex cells: 1-1 and 2-1, 1-1 and 2-2, 1-2 and 2-1, and 1-2 and 2-2. The same is true for this hypothetical organism's mate, who also can produce four different combinations. Collectively, the two mates could produce up to $4 \times 4 = 16$ genetically different organisms through the random process of meiosis. For humans, with 23 pairs of chromosomes, the number is much higher. Given 23 pairs of chromosomes, each mate could produce $2^{23} = 8,388,608$ different combinations, and collectively the two mates could produce $(2^{23})^2 = 70,368,744,177,664$ genetically different offspring.

The process of segregation of chromosome pairs during meiosis can generate a vast amount of genetic diversity. There are, however, some exceptions to the general process that can generate different levels of diversity. Loci are inherited independently only if they reside on different chromosomes or are far apart on the same chromosome. Loci that are close together on a given chromosome are likely to be linked and inherited as a group. In such cases, we refer to different genetic variations as haplotypes, a series of alleles in a set of linked loci.

Linkage does not always occur; during the process of meiosis, recombination of alleles can occur, in which sections from one chromosome are exchanged with the other chromosome. Imagine, for example, a pair of chromosomes with two linked loci (A and B), with alleles *A* and *B* occurring on one chromosome and alleles *a* and *b* on the other chromosome. Because of linkage, we would expect the *A* allele to be inherited with the *B* allele, and the *a* allele to be inherited with the *b* allele. If recombination occurs during meiosis, the sections of DNA can cross over from one chromosome to the other, resulting in a chromosome with alleles *A* and *b* and the other chromosome with alleles *a* and *B*.

2.3 MEASURES OF GENETIC VARIATION

With this brief review of basic genetics, it is useful to examine the kinds of genetic traits used by anthropologists to study human diversity and evolution. Some traits are measured at the level of the phenotype, whereas others are measured in terms of genotype or even the actual nucleotide sequence of DNA. Some examples are given here to illustrate the different ways in which scientists can detect and analyze biological diversity. Our primary interest is traits that show polymorphisms (literally "different forms").

2.3.1 Red Blood Cell Polymorphisms

Many of the traits used by anthropologists to reconstruct population history are genetic variants observable in the biochemistry of red blood cells, including blood groups, blood proteins, and blood enzymes. When you hear someone talk about his or her blood type, he or she is referring to one or more blood groups systems, all defined on the basis of the types of molecules present on the surface of the red blood cells. Red blood cell groups are identified by observing the reaction between antibodies and antigens (substances that produce a reaction to a given antibody). The reaction of someone's blood to different antibodies produces clumping of the red blood cells, allowing different phenotypes to be identified.

The first blood group discovered (in 1900) was the ABO blood group. There are three primary alleles for this system—*A*, *B*, and *O*—as well as subtypes of *A* and some rare alleles. The *A* allele codes for the presence of type-A antigens, the *B* allele codes for the presence of type-B antigen, and the *O* allele codes for neither. The *O* allele is recessive, and the *A* and *B* alleles are codominant. If a person has either the *AA* or *AO* genotype he or she will have type-A antigen, and if a person has either the *BB* or *BO* genotype he or she will have type-B antigen. People with two recessive *O* alleles (genotype *OO*) will have neither antigen present in their blood. People with the genotype *AB* will show *both* type-A and type-B antigens because the alleles are codominant. A sample of a person's blood is tested against two antibodies, anti-A and anti-B, and the reactions are observed. If a person reacts to anti-A but not anti-B, this means he or she has the type-A antigen present but not the type-B antigen, and therefore the phenotype is blood type A. Note that there is no way to determine whether the person has the genotype *AA* or *AO*, because both produce the same phenotype. A person whose blood shows no reaction to anti-A but reacts to anti-B has type-B antigens, and therefore the phenotype is blood type B. If a person's blood shows a reaction to both anti-A and anti-B antibodies, she has blood type AB. A lack of reaction to either anti-A or anti-B indicates blood type O. Because the *O* allele is recessive, there are six different genotypes but only four different phenotypes, and we cannot always tell an individual's genotype.

The ABO blood group is only one of many different types of blood group systems. Another well-known blood system is the Rhesus blood group with a complicated pattern of inheritance involving three linked loci. One simpler way of looking at this blood group involves one of these loci, which has two alleles, *D* and *d*, that determine one's reaction to anti-D antibody. The *D* allele is dominant; people with the genotypes *DD* or *Dd* show a positive reaction to anti-D, and people with the genotype *dd* show a negative reaction. The phenotypes are thus described as "Rh positive" and "Rh negative." The Rhesus blood group system is independent of the ABO system (indeed, these loci are on separate chromosomes, with ABO on chromosome 9 and Rhesus on chromosome 1). When someone says that he or she has "blood type A-positive" he or she is actually using a shorthand code referring to the two separate systems; in this case, the person has phenotype A for the ABO system and phenotype Rh positive for the Rhesus system.

Although the ABO and Rhesus systems are widely known because of medical reasons (e.g., the need for matching in blood transfusion), there are many other blood group systems routinely used by anthropologists to describe patterns of human variation. These systems include the MN, Diego, Duffy, Lutheran, P, Lewis, and Kidd blood group systems among many others. Some of these systems have codominant alleles that allow us a closer look at genetic similarity as revealed by a person's genotype. Loci with dominant alleles are often less useful for certain applications because people with two different genotypes can have the same phenotype. The ability to observe a person's genotype directly moves the analysis of genetic similarities closer to the actual underlying genetic code.

By the 1960s, new technologies were providing more ways of studying genetic diversity. Many new loci were discovered based on proteins in human blood. There are several dozen widely used blood protein and enzyme loci. Individual genotypes were detected in many cases using electrophoresis, a method in which a blood sample is placed into a gel that has an electric charge applied to it; as the electrons move through the gel from negative to positive poles, different proteins move at different rates. By looking at differences in relative movement, a person's genotype can be detected.

2.3.2 DNA Markers

Red blood cell polymorphisms, along with variants of white blood cells and other traits, provide a large number of loci available to study human diversity. Depending on the specific trait, there may be different levels of variation capable of being detected. For loci with dominant alleles, such as the ABO blood group, analysis is confined to observable phenotypes and derived allele frequencies. For loci with codominant alleles, resolution is a bit more specific because we can detect genotypic differences. All of these traits, often collectively referred to as "classical genetic markers," have provided much data on human variation from across the world. New technologies have allowed a closer investigation of genetics through the development of many DNA markers where the focus of analysis is on the underlying genetic code, specifically the actual sequence of nucleotides. Some of these methods are reviewed briefly here.[2]

The first type of DNA marker to be developed was the restriction fragment length polymorphism (RFLP). Certain enzymes, known as restriction endonucleases, cut DNA molecules at particular sequences. For example, the restriction endonuclease known as *Eco*RI recognizes the six-base pair sequence GAATTC and will cut the DNA into fragments where it finds this strand. Mutations in the DNA sequences (such as the substitution of one nucleotide for another or an inserted or deleted section) can change the sequence so that the restriction endonuclease does not cut the fragment. The result, depending on the particular DNA sequence, is fragments of different length, which can be separated using electrophoresis and assigned to different genotypes. As an example, consider a strand of DNA that is eight kilobases long (a kilobase is 1000 bp in length and is abbreviated as "kb"). If we have a restriction endonuclease that recognizes the site at 5 kb into the fragment, then the DNA strand will be cut into two fragments, one 5 kb long and the other 3 kb long. Two alleles can be defined based on whether or not the DNA strand contains the site corresponding to the restriction endonuclease. Let us refer to allele number 1 as lacking the site (producing a single 8-kb fragment) and allele number 2 as having the site (producing a 5-kb fragment and a 3-kb fragment). Three genotypes are possible: 1-1 (two of allele number 1), 1-2 (one of each allele), and 2-2 (two of allele number 2). Depending on the genotype, one or more fragments would be observed. If we saw a single 8-kb fragment, this corresponds to the genotype 1-1, because only 8-kb fragments could be produced by allele number 1. If we found a 5-kb and a 3-kb fragment, this would indicate the genotype 2-2, because allele number 2 results in two fragments of different length. The genotype 1-2 has a copy of both alleles and would produce three fragments of 3 kb, 5 kb, and 8 kb in length. RFLPs can be considered as codominant loci, but they reveal the actual genetic code more closely than blood groups or other classical markers.

Another type of DNA marker focuses on variation in satellite DNA, sections of noncoding DNA that contain a number of tandem repeats of nucleotides. Because of mutation, different chromosomes will contain variation in the number of times a given section is repeated. For example, the sequence TCT**GAGAGA**GGC has three repetitions of the dinucleotide sequence GA, shown here in boldface. The sequence TCT**GAGAGAGAGA**GGC has five repetitions of the sequence GA. Alleles are defined by the number of repeats. Electrophoresis can sort DNA fragments by size, allowing different alleles to be identified. Microsatellite DNA (also known as short tandem repeats, or STRs) consists of repeated segments of between one and five nucleotides. Minisatellite DNA (also known as variable number of tandem repeats, or VNTRs) consists of repeated segments greater than five nucleotides.

Some sections of DNA are repeated and interspersed throughout the genome rather than at a single location. *Alu* insertion polymorphisms are one example, unique to primates. A section of DNA of roughly 300 bp in length has been found repeated throughout the entire genome. In humans, there are approximately half a million copies collectively accounting for about five percent of our genome. Some of these polymorphisms are found only in humans and are absent in our closest living relatives, the African apes. This property is useful because we are thus able to define the ancestral state of the polymorphisms.

A rapidly developing form of DNA analysis focuses on single-nucleotide polymorphisms (SNPs) that refer to specific positions in the DNA where there are two alternative bases that each occur in at least one percent of the population. For example, the following two sequences differ at one site, indicated in boldface. The first strand has the base A and the second has the base C.

GAATTAGTCAAGCAGGTC**A**GATACTATTGTCTGCT

GAATTAGTCAAGCAGGTC**C**GATACTATTGTCTGCT

Estimates suggest that there may be over 200,000 SNPs in human genes and perhaps 10 times that number in noncoding DNA.

All of the above methods focus on identification of small differences in DNA sequences among individuals. An ultimate goal is to be able to compare large DNA sequences. The goal of the Human Genome Project is to identify the complete DNA sequence of the human genome, which is estimated to be approximately three billion base pairs in length. Although still in relative infancy, the field of molecular genetics has vastly increased our knowledge of genetic variation, allowing closer and more specific assessment of genetic variation.

2.3.3   Mitochondrial DNA

The human genome is made up of nuclear DNA and a small amount of DNA that is found within the mitochondria, organelles within the cell responsible for energy production. Mitochondrial DNA (mtDNA) is a circular molecule averaging 16,569 bp in length that has been completely sequenced. Human mitochondrial DNA contains 37 different genes that are linked and inherited as a single unit. Mitochondrial DNA consists of two regions, the coding region containing the genes and a control region that regulates the replication of mtDNA. The coding region is analyzed using RFLPs, combining the RFLP alleles for a set of restriction enzymes to define a specific mitochondrial haplotype. Direct sequencing of DNA is used to study variation in the control region.

The human mitochondrial genome is only about 0.0005% the size of the nuclear genome but has several interesting properties that make it particularly useful for assessing population history. Mitochondrial DNA is maternally inherited; your mitochondrial DNA came from your mother, but not your father, unlike nuclear DNA, which is inherited from both parents. Because of inheritance from a single parent there is no recombination, and analysis of ancestral connections is made easier. You carry your mother's mitochondrial DNA, she in turn carries her mother's, and so forth back into the past. In addition, mitochondrial DNA has a relatively rapid rate of mutation, allowing evolutionary inferences about more recent events. As will be seen in Chapter 4, the analysis of mitochondrial DNA is central to much of the debate over modern human origins.

2.3.4   The Y Chromosome

One of our 23 pairs of chromosomes are the sex chromosomes, which come in two forms, X and Y, and determine one's sex—females typically have two X chromosomes (XX) and males have one X and one Y chromosome (XY). Because a female will always pass on an X chromosome, the genetic contribution of the male (X or Y) determines the sex of the offspring. Because the Y chromosome is smaller than the X chromosome, males have a number of loci that are on their X chromosome that are not represented on their Y chromosome, and the gene on the X chromosome is expressed by itself, a condition known as an X-linked trait. The lack of a physical counterpart on the Y chromosome explains why X-linked traits such as hemophilia are more common in males than females. The hemophilia allele is recessive, which means a female needs to inherit two copies to manifest the disease. A male need only inherit the allele for hemophilia from his mother and he will manifest the disease. The probability of any male having hemophilia is equal to the frequency of the allele in the population, whereas the probability of a female having hemophilia is equal to square of this probability. As an analogy, consider that the probability of flipping a coin and getting "heads" is equal to 0.5, whereas the probability of getting *two* heads is $0.5^2 = 0.25$.

Except for a small piece at each end, most of the Y chromosome is nonrecombining and is inherited solely from the father. The nonrecombining portion of the Y chromosome (NRPY) of a male is inherited from his father, who inherited it from his father, and so on. Like mitochondrial DNA, analysis of Y chromosome polymorphisms provides an easier way to reconstruct population history without the complications introduced by two parents, multiple ancestors, and recombination. Y chromosome polymorphisms have been detected using RFLP analysis, satellite DNA analysis, and direct sequencing.

2.3.5 Complex Phenotypic Traits

Before the development of DNA-based data or even classical genetic markers, anthropologists relied on the measurement and analysis of physical traits, such as height, skin color, and cranial length to quantify variation and biological similarity. Analysis of complex phenotypic traits is complicated by the fact that we often know little about the relationship between genotype and phenotype and that many of these traits can be affected by developmental changes throughout one's lifetime. The phenotype is not simply a reflection of the mode of inheritance but also represents an environmental component. Two people with identical skin color could have different genotypes because of differing combinations of genes producing the same effect and because of environmental differences. The fact that some traits, such as human height, change so quickly from one generation to the next is ample evidence that many complex phenotypic traits are extremely "plastic."

Still, complex phenotypic traits are in part a reflection of one's genetics and thus have the potential, with appropriate methods and controls, to yield information about genetic relationships and population history. In addition, they are often the only major source of data available for populations in earlier times, either during the nineteenth century or deep into the fossil record. The ability to determine blood types and DNA sequences from ancient material is useful, but for much of our past the primary source of information is from complex traits. Moreover, as shown later, we can often use complex trait variation in living humans to answer questions about population history, much in the same way as we use classical genetic markers and DNA markers.

Several types of complex traits have routinely been studied by anthropologists. Anthropometrics are measurements taken of the body, head, and face of living humans. Body measures include measurements of height, weight, fat distribution, and a variety of limb lengths and body breadths and circumferences. Measurements of the head and face, well known to be quite useful in reconstructing population history, include lengths and widths across the skull and face. The same measurement methods are also applied to skeletal data from the body (osteometrics) and the skull and face (craniometrics). Other measurements of complex traits include skin color, dental measures, and fingerprint patterns.

2.4 HOW DOES EVOLUTION WORK?

The ultimate outcome of measuring human genetic variation is a set of frequencies of different alleles for genes and DNA sequences (or, in the case of complex traits, a series of average measurements that are related to the underlying allele frequencies). Our primary concern is how allele frequencies change over time. The process of evolution is best described first in terms of the changes in the relative frequencies of alleles from one generation to the next, a process known as microevolution.[3] The study of microevolution shows us how genetic change can occur in populations over relatively short intervals of time. The same process can then be extrapolated to longer intervals, including the origin of new species.

The study of microevolution focuses on the relative frequencies of different alleles of genes or DNA sequences. An allele frequency is simply the proportion of all alleles in a population. For example, imagine a population of 100 people that have been typed for the MN blood group, in which we find 36 people with genotype *MM*, 48 people with genotype *MN*, and 16 people with genotype *NN*. We are interested in the relative frequency of the *M* and *N* alleles. The first thing we do is to count the number of *M* alleles. Anyone with the genotype *MM* or *MN* has at least one *M* allele. Everyone with the genotype *MM* has two *M* alleles, and everyone with the genotype *MN* has one *M* allele. Remember to count everyone with genotype *MM* twice because they each have two *M* alleles. In this example, the total number of *M* alleles is 2(36) + 48 = 120 *M* alleles. We now do the same thing with *N* alleles. Because there are 48 people with one *N* allele (genotype *MN*) and 16 people with two *N* alleles (genotype *NN*), the total number of *N* alleles is 48 + 2(16) = 80. The total number of alleles (*M* and *N*) is 120 + 80 = 200. This number is twice the number of people in the population, which is expected because each person has two alleles.

The relative frequency of the *M* allele is computed as 120/200 = 0.6. Likewise, the relative frequency of the *N* allele is 80/200 = 0.4. We therefore would characterize this population and locus as having allele frequencies of 0.6 *M* and 0.4 *N*. Our interest is now in how these frequencies would change. Under ideal conditions, known as Hardy–Weinberg equilibrium, these allele frequencies would remain the same from one generation to the next. In the real world, allele frequencies can change from one generation to the next. Microevolution is thus defined as a change in allele frequencies over time. Population geneticists have worked out the different ways in which microevolution can occur. There are four evolutionary forces that cause changes in allele frequencies over time: 1. mutation, 2. natural selection, 3. genetic drift, and 4. gene flow.

2.4.1 Mutation

Mutation is a random change in the genetic code. Mutations occur all the time in nature for a variety of reasons, including background cosmic radiation. Mutations are the ultimate source of genetic variation; without mutations, there would be no variation and hence no evolution. There are a number of different types of mutations. Some mutations consist of a single change in the genetic code, where one nucleotide is substituted for another, such as "C" mutating into "G." Other mutations include deletions and insertions of nucleotides, reversals of entire sections of DNA, repeated sections of DNA, and duplication of entire chromosomes, among others. Although this process occurs in all cells, a mutation must occur in a sex cell to be passed on to the next generation and have an evolutionary impact.

If we view the process of reproduction and inheritance as the transmission of information from one generation to the next, mutations represent errors in the transmitted information. Mutations change the relative proportions of alleles and cause a small amount of microevolution. Imagine a locus with one allele, A, in a population of 25 couples (for a total of 50 people) If everyone in the population has two copies of the A allele (genotype AA), then the frequency of A is $100/100 = 1.0$. If each couple has two children, they will each pass on two A alleles on to the next generation. Now, imagine that there is a mutation in one of the sex cells, causing a change from the A allele into a new allele, which we will call B. The next generation will have 99 A alleles and one B allele. The frequency of the A allele has gone down from 1.0 to $99/100 = 0.99$ and the frequency of the B allele has increased from 0 (it wasn't there to begin with) to $1/100 = 0.01$.

The important function of mutation is to introduce something *new* into a population. In the above example, a new allele (B) has been introduced. Whether the frequency of B increases, decreases, or remains the same depends on the other evolutionary forces. We are most interested in the forces that lead to relatively high values for both alleles. We use the term genetic polymorphism to refer to a locus where there are at least two alleles that have frequencies greater than 0.01. Genetic polymorphisms represent cases in which mutation alone cannot account for the observed variation and the other evolutionary forces are responsible. The trick is figuring out which one(s).

2.4.2 Natural Selection

Charles Darwin and Alfred Russel Wallace independently developed the concept of natural selection. The basic idea is fairly simple, particularly if we start by considering the kind of artificial selection that humans have applied to the domestication of plants and animals for thousands of years. Imagine, for example, that you have just inherited a pig farm. You plan to sell some pigs at the market and to keep some for future breeding stock. You will notice that not all pigs are the same—there will be variation in size and shape as well as temperament. You are particularly interested in the variation in body size, noting that some pigs are large and plump, and other pigs are small and scrawny. You want to raise the largest, fattest pigs possible. What do you do? Which pigs go to market, and which remain behind as breeders for the next generation of pigs? The basic principle is to take the pigs that have the characteristic you desire and keep them for breeding. You will select the fattest males and females, noting that if fatness is at least partially inherited, then you are likely to produce future generations that will become, bit by bit, slightly fatter on average each generation. You are *selecting* which individuals contribute to the next generation and which ones are sold (or wind up as pork chops).

Darwin noted that a similar process could also occur in nature, where an organism's probability of surviving and reproducing is in part a function of how well adapted it is to a given local environment. For example, finches with stronger beaks are more likely to be able to crack open the hard seeds remaining after a drought and are thus more likely to pass on any genes affecting beak size to the next generation. Each generation will show variation around the average beak

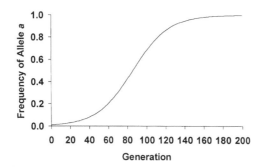

Figure 2.3. *Simulation of natural selection. This simulation uses a simple locus with two alleles, A and a. The initial frequency of the a allele is set to 0.01. Fitness values for the three genotypes are set to favor individuals with one or more a alleles: AA = 90%, Aa = 95%, aa = 100%. Because individuals with the aa genotype are the most fit and individuals with the Aa genotype the next most fit, this simulation favors an increase in the frequency of the a allele over time. As a consequence, the a allele changes from a very low frequency (0.01) to close to 1.0 in a short period of time.*

size, but over time, selection for larger beaks will lead to an increase in average size. Because this process of selection occurs as a function of nature, we refer to it as *natural* selection.

The effect of natural selection on allele frequencies depends on whether a trait is favored in a particular local environment, and it is easily modeled. Figure 2.3 illustrates an example of selection using a hypothetical locus with two alleles, *A* and *a*. In this case, I set the fitness (the probability of survival and reproduction) equal to 90% for genotype *AA*, 95% for genotype *Aa*, and 100% for genotype *aa*. This simulation is an example of a situation in which individuals with two *a* alleles are the most fit, individuals with only one *a* allele are somewhat less fit, and individuals with no *a* alleles are the least fit. Thus, for every 100 individuals with genotype *aa* that survive and reproduce, only 95 with genotype *Aa* and 90 with genotype *AA* survive and reproduce. I started the simulation with a low initial frequency of the advantageous allele *a* (= 0.01). The initial change in the frequency of the *a* allele is relatively slow, reaching a value of only 0.029 after 20 generations. The rate of change increases rapidly after 60 generations as the frequency of *aa* individuals increases. After 100 generations, the frequency of *a* has reached a value of 0.68, and it is almost 0.97 after 150 generations. After 200 generations, the frequency of the *a* allele has increased to 0.997. This example shows how fast natural selection can act—the *A* allele has been virtually replaced by the *a* allele after only 200 generations, a short time in evolutionary terms. The speed of natural selection depends on the initial allele frequencies and the fitness values for each genotype. If the differences in fitness are greater than the above example, selection will occur more rapidly, and if the differences in fitness are less, then selection will be slower. It is obvious, however, that even a small difference in fitness can add up over a relatively short time.

The example in Fig. 2.3 represents a typical pattern of selection *for* one allele and *against* the other. The eventual outcome is the replacement of one allele by another. There is also a form of selection where the heterozygote is the most fit, which leads to a balance in allele frequencies. Figure 2.4 illustrates an example of balancing

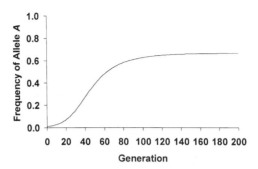

Figure 2.4. *Simulation of balancing selection where the heterozygote has the highest fitness. This simulation uses a simple locus with two alleles, A and a. The initial frequency of the A allele is set to 0.01. Fitness values are set to favor individuals who are heterozygous (genotype Aa): AA = 95%, Aa = 100%, aa = 90%. The frequency of the A allele increases slowly at first, and then more rapidly after 30 generations, reaching an equilibrium value after about 80 generations. The A allele does not replace the a allele, because fitness is maximized at an intermediate value when the heterozygote is the most fit genotype.*

selection. Here, the fitness for genotype AA is 95%, the fitness of genotype Aa is 100%, and the fitness of genotype aa is 90%. Genotype Aa (the heterozygote) is the most fit, followed by AA and then by aa. The outcome of this form of selection is less clear at first; because aa has the lowest fitness, we might expect that allele a will be replaced by allele A. However, because the heterozygote is the most fit, *both A and a alleles are put back into the population.* Figure 2.4 shows the typical pattern of evolution through selection for the heterozygote. The simulation starts with a low frequency of A (= 0.01) that increases slowly at first and begins leveling off after about 80 generations. Note that the A does not completely replace a—the final frequencies represent a balance.

An excellent example of selection for the heterozygote in humans is the case of the sickle cell allele (S), which is a mutation in of the normal allele (A) of the gene responsible for the beta chain of the hemoglobin protein. Individuals with the genotype AA have normal hemoglobin that transports oxygen efficiently. People with the genotype SS have the genetic disease sickle cell anemia, in which the blood cells can distort under stress, thus reducing oxygen transfer and often leading to death. Heterozygotes (AS) carry the harmful gene but are affected only slightly if at all. In many environments, the S allele is selected against because of the very low fitness of people with the SS genotype. In environments with epidemic malaria, the situation is different, because those with genotype AS have resistance to malaria and are the most fit. The balance in selective forces (deaths from malaria and deaths from sickle cell anemia) has led to frequencies of S as high as 0.2 in human populations where malaria is epidemic.

2.4.3 Genetic Drift

Genetic drift is the random change in allele frequency from one generation to the next because of basic probability. In other words, allele frequencies may

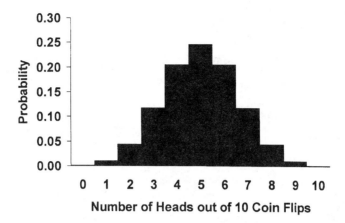

Figure 2.5. *Probability distribution for 10 coin flips. This graph represents the probability of getting a given number of heads from 10 coin flips. The most likely outcome is five heads and five tails, but this will only happen about 25% of the time. The rest of the time there will be an uneven number of heads and tails.*

increase, decrease, or stay the same, all as the result of chance events in reproduction. Perhaps the simplest way to understand genetic drift is to consider the analogy of flipping a coin. Whenever you flip a coin, there is a 50:50 chance of getting a head or a tail. If you flip 10 coins, you might therefore expect to get 5 heads and 5 tails. Although this is one possible outcome, you might also get 6 heads and 4 tails, 2 heads and 8 tails, or any other combination adding up to 10 coin flips. If you continued flipping sets of 10 coins repeatedly, you expect *on average* to have 5 heads and 5 tails, but you also know that other possibilities could occur for any individual flip of the coins. We cannot tell beforehand how many heads will result, but we can compute the probability of such an event. Figure 2.5 shows the probability of getting a certain number of heads out of 10 coin flips. The resulting "bell-shaped" curve is typical of this type of probability distribution, showing the greatest probability in the middle (5 out of 10 are heads) and smaller probabilities for a fewer or greater number of heads. Note that most of the time you will *not* get exactly 5 heads.

The same principles of probability and sampling occur in evolution. Suppose, for example, that you are heterozygous for a given locus with a genotype of *Aa*. There is a 50% chance that you will pass on the *A* allele to a given child and a 50% chance that you will pass on the *a* allele. If you have 10 children, you expect that *on average* you will pass on 5 copies of the *A* allele and 5 copies of the *a* allele. As shown in Fig. 2.5, most of the time this will not happen. You might pass on three *A* alleles and seven *a* alleles. In this case, you have contributed more *a* alleles than expected, purely because of chance. This random process is occurring in everyone within a population, so that the allele frequencies are likely to change from one generation to the next. If the parental population has an allele frequency of 0.5, the frequency in the next generation might be the same but also might be lower or higher. Genetic drift is a random process in which allele frequencies could change in either direction, and at different magnitudes, generation to generation. Allele frequencies could continue to change until an allele

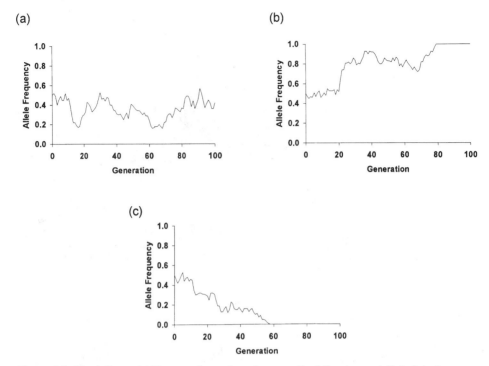

Figure 2.6. *Simulations of 100 generations of random genetic drift using an initial allele frequency of 0.5 and a population size of 50 reproductive adults. Each simulation of genetic drift will be different.*

reaches extinction (all gone) or fixation (everyone has it) by chance. In that case, the allele frequency will remain at 0.0 or 1.0 until a new allele is introduced through mutation or migration.

Three examples of genetic drift are shown in Fig. 2.6. I used a computer simulation to mimic the process of genetic drift by deriving random numbers to determine which of two alleles gets passed on to the next generation. In each case, I started the situation at an allele frequency of 0.5 and let genetic drift occur for 100 generations in a population of 50 people (25 couples), all of who had 2 children each (thus keeping the population size constant, to make things easier for the moment). In the first graph (Fig. 2.6a), the frequency drifts up and down randomly and after 100 generations has a value (0.42) not too much different from where it started. Figure 2.6b shows a case where drift continues, up and down, until by chance the allele reaches fixation. At this point, everyone in the population has two copies of the same allele—the other allele has been lost. Figure 2.6c illustrates a similar example, although in this case the allele frequency drifts until the allele is completely lost.

Given enough time, all populations will eventually drift toward extinction or fixation of an allele. Some populations will drift more than others because of differences in population size. The magnitude of genetic drift in any generation is greater in smaller populations and less in larger populations. Although there are different

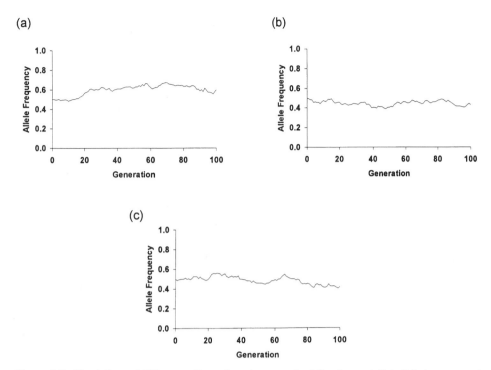

Figure 2.7. *Simulations of 100 generations of random genetic drift using an initial allele frequency of 0.5 and a population size of 1000 reproductive adults. Notice that there is much less genetic drift than is seen in Fig. 2.6, where the population size was 50. The magnitude of drift varies inversely with population size.*

measures of population size that will be discussed later (Chapter 7), the measure referred to here is the number of reproducing adults in a population. Figure 2.7 shows the process of genetic drift over 100 generations in three populations where the population size is 1000. Drift still occurs over time, but the magnitude of drift in any generation is much less than shown in Fig. 2.6, which is based on a much smaller population size. The fact that larger populations drift less is expected from basic probability theory—the larger the number of events, the less deviation from the expected numbers. It is no surprise to flip a coin four times and get all heads, but you would be very unlikely to flip 4000 coins and get all heads!

It is important to remember two points when considering the potential impact of genetic drift in human evolution. First, all populations will show some effect of drift, and even in large groups the cumulative effect over many hundreds of generations can be large (although the actual pattern of change is also affected by the other evolutionary forces). Second, human populations have been small throughout most of our species' history. Hunters and gatherers frequently live in small local bands of about 25 individuals that are part of larger tribal units numbering about 500.[4] The actual number of reproductive individuals in any generation is less, roughly half the total number. There would have been much potential for genetic drift in our ancestry.

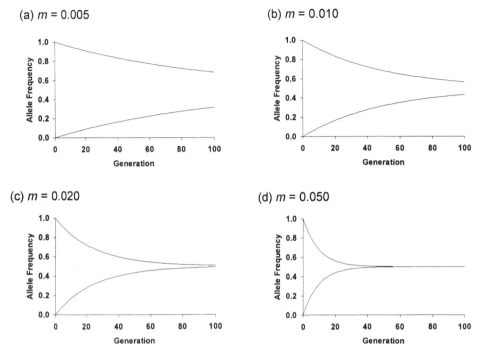

Figure 2.8. *Simulations of 100 generations of gene flow between two populations. In each simulation the initial allele frequencies were set to 1.0 and 0.0 for the two populations. Each generation, a proportion m of each population exchanges genes with the other population. All four graphs show how gene flow acts to make populations more genetically similar over time. The higher the rate of gene flow, the more quickly they become similar.*

2.4.4 Gene Flow

Gene flow occurs when an individual from one population moves into another population and reproduces there. Gene flow has two major effects on allele frequencies. First, new genes can be introduced into a population. Second, gene flow between a pair of populations makes them increasingly more similar genetically over time because of the mixing of genes. This effect is illustrated in Fig. 2.8, which examines the effect of gene flow on the allele frequencies in two populations. At the start of the simulation, the allele frequency is 1.0 in one population and 0.0 in the other population, the maximal difference possible. Each generation, gene flow was simulated by allowing a small proportion (m) of the individuals to move to the other population (thus leaving the proportion $1 - m$ behind). Different amounts of gene flow were used, ranging from a per-generation rate of $m = 0.005$ to a rate of $m = 0.050$. In all cases, the two populations become increasingly similar over time. These graphs also show that the higher the rate of gene flow, the more quickly the two populations resemble one another.

2.4.5 Interaction of the Evolutionary Forces

For simplicity of explanation, the four evolutionary forces have been discussed separately. In reality, all four can be operating at the same time, interacting to produce complex patterns of allele frequency change over time both within and between populations. The fate of a mutant allele is affected by the interaction of natural selection, genetic drift, and gene flow. If the mutant is harmful, natural selection will reduce its frequency, but if it is an advantageous mutation, then selection will lead to an increase in frequency. Genetic drift will most often tend to eliminate a new mutant, because the frequency of a new mutant is very low and thus more easily lost in a subsequent generation. However, in some cases the frequency of a mutant allele will increase randomly because of drift, even in some cases where the allele is selected against. Gene flow allows the opportunity for a new mutant allele to be transmitted into another population.

The evolutionary forces sometimes oppose one another and sometimes act in concert. To understand this concept, consider how the evolutionary forces affect the level of genetic diversity *within* a population. By adding something new, mutation can increase diversity slightly. Because genetic drift acts, over time, to fix or eliminate an allele, the average impact of genetic drift is to reduce diversity within a population. Because genetic drift is affected by population size, we expect on average to see lower levels of genetic diversity in small populations, a topic returned to in Chapter 5. Gene flow, on the other hand, can increase diversity within a population because it introduces new alleles. A balance between mutation rate, population size, and the rate of gene flow determines the evolution of diversity in a population. It is not always easy to determine the relative roles of the evolutionary forces. If, for example, you compare two populations and find higher levels of genetic diversity in one of them, is this because of larger population size, higher rates of gene flow, or both?

Natural selection can also affect the level of genetic diversity in a population, depending on the type of selection and the initial allele frequencies. If there is selection for one allele over another, then the frequency of one allele will increase at the expense of the other and diversity will be reduced. If, however, there is selection for the heterozygote, then diversity could increase (for example, as in Fig. 2.4).

When considering recent human evolution we are also interested in changes in the genetic relationships *between* different populations. The evolutionary forces can act to increase diversity between populations (that is, make them more different) or decrease variation between populations (make them more similar). Genetic drift will tend, on average, to make populations more dissimilar over time, whereas gene flow will tend to make them more similar. Natural selection could increase or decrease genetic differences between populations depending on the type of selection and whether the populations are in different environments with different optimal adaptations.

2.4.6 Microevolution and Macroevolution

The study of microevolution looks at changes in allele frequency over relatively short intervals of time. When considering the long-term evolution (macroevolution) of a group of organisms, such as humans and their ancestors, we must consider the cumulative effect of evolutionary change, particularly its effect on the origin of a species.[5] The term "species" has a number of different uses and definitions. A common definition, the biological species concept, focuses on reproductive capability. If organisms from two populations naturally interbreed and produce fertile offspring, they are considered in the same species. If these conditions are not meant, they are placed in different species. Consideration of different species reveals that some species are more closely related than others are. Organisms such as mosquitoes and elephants are clearly different species that have accumulated many genetic differences over the course of hundreds of millions of years. Other organisms, such as the horse and the donkey, are much more similar both physically and genetically and are similar enough that they can produce offspring (known as mules). However, mules are sterile, indicating that horses and donkeys are reproductively isolated evolutionary units no longer capable of sharing their genes in future generations.

Although the biological species concept is in theory practical for classifying living organisms, it falls somewhat short when considering fossils, where there is no direct evidence on viable mating and species differences must be inferred from physical differences. This problem has been compounded in the past by the use of species names as descriptive labels separating members of an evolutionary lineage over time. Thus earlier populations of humans might be referred to by a different species name, even if they are considered part of a single evolutionary line—we just use the labels to demarcate different stages along the continuum, much as we use labels such as "infant" and "child" to note different stages along an individual's life cycle. To many evolutionary biologists, the use of species names as labels is inappropriate; they argue that such names should be given only to reproductively separate groups.

We return to the issue of species names (in reference to the human fossil record) in the next chapter. For the moment, long-term evolution is considered in two ways: 1. Anagenesis, the evolution of a single evolutionary lineage (whether called a single species or separated arbitrarily using species names as labels), and 2. Cladogenesis, the origin of new species through a "branching" process.

The mechanism of anagenesis can be considered in terms of extrapolations of the evolutionary forces over time. Here, a species is divided into a number of populations all connected by gene flow. New mutations emerge and are shared throughout a species through gene flow. There is a changing dynamic between genetic drift and gene flow, the former acting to increase populational differences and the latter reducing them. Selection will sometimes reduce populational differences and other times cause a geographic gradient related to environmental variation. The important point in discussing anagenesis is that the species maintains its unity through gene flow but at the same time allows local and regional variations to be produced and maintained.

Anagenesis deals with a single species. This species evolves through the interaction of the evolutionary forces and may in fact change sufficiently that we would refer to different stages as different species for labeling purposes, but it is a single reproductive entity at any one point in time. Many people equate evolution in general with the specific outcome of anagenesis—earlier fish evolved into more recent fish, earlier horses into more recent horses, and earlier humans into more recent humans. Taking anagenesis to be the whole of evolution gives us a distorted image of the fossil record, with each living species stretching back in time to an earlier version of itself. The equation of evolution with anagenesis alone also gives rise to a very popular misconception regarding human evolution. This misconception is shown in a typical argument that I have heard on a number of occasions, which goes like this: "You anthropologists claim that apes evolved into humans. If this is true, then why do we have both apes and humans today? If the apes turned into humans, they shouldn't be around anymore. The fact that they are alive in the world today is proof that we didn't evolve from apes."

Although this argument may sound reasonable, its logic is twisted by a misconception of macroevolution that takes *all* of long-term change to be solely the result of anagenetic change. I have found an analogy that sometimes clarifies this misconception. Consider the fact that any one of us has two direct ancestors, our biological mother and father. In a genetic and evolutionary sense, we came from them. Did they disappear when you came into being? Of course not. Or did the older generation and the new one live side by side? It might be argued that this analogy is not really the same thing, because our parents do not turn into us but rather give birth to us, a process that is conceptually different from anagenesis. True, but this is not different from cladogenesis, the birth of a new species. Just as your parents gave rise to a new branch (you), a parental species can give rise to a new daughter species without necessarily disappearing right away. Cladogenesis views macroevolution as a family tree with many different branches, whereas anagenesis is more like a tree with a single trunk. Adding to the analogy of a life cycle, species also die out; in fact, over 99% of all recorded fossil species are now extinct. Cladogenesis and extinction represent the birth and death of species and provide a turnover in life, similar to births and deaths within our own species. No human who lived two centuries ago is still alive today, yet our species has not died out. The new replace the old in a constant balance between birth and death.

How then does speciation occur? For a new species to be "born" two things must happen. First, some part of the original species (usually a small population) must become reproductively isolated; that is, gene flow must be cut (or at least significantly reduced). Because gene flow is the "glue" holding a species together, its impact must be lessened by some part of the parent species becoming reproductively isolated. Reproductive isolation can result from a number of factors, although geographic separation limiting gene flow appears to be one of the most common. If populations become geographically distant or isolated, they are less likely to share genes with other populations. Reproductive isolation is a necessary, though not sufficient, factor in speciation. Genetic divergence between populations must then take place. Mutations appearing in one population may not appear in others. Genetic drift will act to make an isolated population genetically different, particularly if the isolated group is small. Natural selection can also have an impact, especially if the isolated population is in a different environment, or otherwise experiences different adaptive pressures, than the parent species. The interaction of evolutionary forces can produce situations in which the daughter population becomes increasingly different genetically (evolutionary biologists continue to debate the nature of this interaction in different groups of organisms). If enough time goes by, and enough genetic divergence takes place between the parent and daughter population, then they will become different enough that they are no longer capable of producing fertile offspring.

2.5 GENETICS AND HISTORY—TWO EXAMPLES

In a broad sense, reconstructing history from genetics can be done from two perspectives, corresponding to whether our interest is the reconstruction of the history of populations *within* a single species over time or the history of relationships *between* species over time. Both approaches focus on genetic similarity as a clue to evolutionary history—the more similar two samples are, the more likely they are to be related. The difference is in the interpretation of past events—did evolutionary change take place within a species or between species?

In studies of population history, the focus is on genetic relationships that reflect the degree of common ancestry of the units being studied, be they different populations within a species or different species. Because natural selection can distort such relationships, making groups appear more or less related than they really are, we tend to view the effects of selection as "noise" that must be removed to discover the "signal" of population history. One way of dealing with this problem is to look only at traits that are neutral—they have not been appreciably affected by natural selection. The problem here is knowing which traits are neutral, something that is not always clear. We also need to deal with sampling error among traits, so that we are not getting a distorted picture from a single, less typical trait. Here, we would average measures of genetic relationships over as many traits as possible. None of these solutions is perfect, but verification with different data sets often provides a reasonable solution.

Two case studies are presented here to illustrate some of the different approaches to the study of genetics and population history. The first looks at recent evolutionary trends within a species—human populations in Ireland. The second looks at evolutionary history of two different species separated by several million years of evolution—humans and chimpanzees.

2.5.1 The Population History of Ireland

The history of the island of Ireland (today made up of two nations, the Republic of Ireland and Northern Ireland) has long been of interest to historians and anthropologists. Ireland was first occupied over 4000 years ago, followed by Viking and Anglo-Norman invasions and Scottish and English immigration. The demographic history of Ireland shows a major increase in population size in the early 1700s after the introduction of the potato and a major decline in population during the Great Famine (1846–1851) when the potato crop failed repeatedly. Changing patterns of population size and migration (both internal as well as from outside of Ireland) are of particular interest because of their impact on patterns of genetic drift and gene flow.

My interest in Irish population biology goes back to graduate school, when I worked on several Irish studies with my graduate advisor, Frank Lees, and his graduate advisor (my "academic grandfather"), Michael Crawford. Of particular interest was a large set of anthropometric data collected by C. Wesley Dupertuis in the mid-1930s as part of an anthropological survey of Ireland conducted by Harvard University.[6] Dupertuis collected anthropometric and demographic data on almost 9000 adult Irish men across the entire island. His colleague Helen Dawson collected an additional sample of almost 2000 adult women. Mike obtained copies of the original coding forms in the early 1970s and began transferring these data into computer form. By 1978, I was in graduate school looking for a dissertation topic and began working on a small subset of these data, focusing on the relationship of geographic distance to biological variation in a small area of the western coast of Ireland. Over the years, Mike and I frequently talked about getting all of the data into computer form for a comprehensive analysis of the population history of the entire island. After a number of years (and intervening projects), I returned to the Irish problem, obtaining a research grant from the National Science Foundation to finish the electronic transfer of the data and begin analysis.

Of course, these analyses deal with anthropometric rather than genetic data. As noted earlier, anthropometric data do reflect genetic differences between populations but are also influenced by environmental differences. Fortunately, in some cases the phenotypic variation can be used to assess the underlying genotypic variation, especially in cases in which the populations are similar environmentally and culturally. Our purpose was to use the *phenetic distances* (measures of overall phenotypic dissimilarity) between populations as a relative measure of the underlying *genetic distances* (measures of overall genetic dissimilarity). John Blangero, Sarah Williams-Blangero, and I have shown in several papers how anthropometric data can be used to estimate genetic distances between populations.[7]

We wound up with complete data on 17 anthropometric measures for over 7000 adult men whose parents were both born in Ireland. Because anthropometric data can be influenced by age differences, we started by statistically removing age-related effects and confined all subsequent analyses to the age-adjusted data. Because body measurements are frequently affected by environmental and dietary influences, we confined one set of analyses to 10 head measures that were shown in other studies to provide a reasonable estimate of genetic affinity between populations. Each individual was placed into his county of birth, which we used as a rough surrogate for a geographically defined population. We then used methods of estimating the phenetic distances from anthropometric data, winding up with a matrix of distances that contained 31 rows and 31 columns corresponding to the 31 birth counties (actually, there are 32 counties in Ireland—26 in the Republic of Ireland and six in Northern Ireland—but we lacked adequate data for one county, bringing our total to 31). This matrix is incredibly long—it contains the phenetic distance between each pair of populations; county 1 is compared to county 2, county 3, and so forth, for a total of 465 distances. You can look at such matrices all day long without seeing much pattern, because there are too many numbers to juggle in your mind. Instead, we reduced this large matrix into a single simple picture using a data reduction method known as principal coordinates analysis.

Figure 2.9 shows the results in the form of a "phenetic distance map," a graphic representation of the phenetic similarity of populations; the closer two groups are together on the map, the more phenetically similar they are. It is clear that four of the counties (indicated as triangles in the plot) are separated from the rest of the counties. What do these counties have in common that might explain their relative similarity to each other and their difference from the remaining Irish counties? All four counties are located together in the Irish midlands. We might expect that geo-

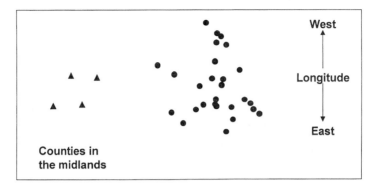

Figure 2.9. *Phenetic distance map of 31 counties in Ireland (Republic of Ireland and Northern Ireland) based on 10 head measures. This "map" provides a graphic representation of biological similarity; the closer two points on the map, the more similar the populations. Two patterns are apparent in this figure: the genetic distinctiveness of the Irish midlands and the west-east difference. Adapted from Relethford and Crawford (1995).*

graphically central populations would actually be similar to all other counties, because most migration events would cross the center of the island, but this is not the case. The common factor here is history—the Irish midlands appear to have experienced greater than average Viking admixture. Viking invasions of Ireland began in 794 and were often confined to coastal populations, such as Dublin. However, Vikings also sailed from the Atlantic Ocean up the Shannon River into the Irish midlands, where they had several major settlements. The large number of invaders, combined with subsequent mating, appears to have introduced a larger amount of Viking genetic admixture than elsewhere in Ireland, thus making these counties slightly different.

The phenetic distance map shown in Fig. 2.9 also revealed another, less immediately obvious, pattern. When we looked at the geographic locations of the other Irish counties (the dots in Fig. 2.9), we noticed a strong correlation with longitude—the west coast counties plotted at the top of the graph and the east coast counties toward the bottom. This strong pattern agrees with other studies of Irish population genetics, including analysis of classical genetic markers, which show a basic east-west longitudinal gradient. We interpreted this gradient as a reflection of differences in immigration history, because the plantation of settlers from England and Wales that started in the early 1600s occurred more often in the north, east, and southeast of Ireland. Our conclusion was that a combination of immigrations, with differential gene flow, is the primary factor affecting the recent genetic history of Ireland. We tested our hypotheses further by comparing our Irish data with published anthropometric data from England and from Norway and Denmark, the latter two indicative of the major sources of Viking admixture. Figure 2.10 shows the resulting phenetic distance map comparing six geographic regions within Ireland to England, Norway, and Denmark. As expected, the Irish midlands plot closest to Norway and Denmark, confirming our hypothesis that the midlands are most similar to these Scandinavian populations as the result of Viking admixture. Figure 2.10 also shows the longi-

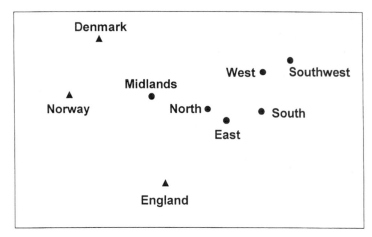

Figure 2.10. Phenetic distance map comparing six geographic regions in Ireland with Denmark, Norway, and England, based on head measures. The Irish midlands are the most similar to Denmark and Norway, whereas eastern counties are more similar to England. Adapted from Relethford and Crawford (1995).

tudinal gradient, with the northern, eastern, and southern regions of Ireland being most similar to England. Recent work by Kari North and colleagues has confirmed these historical influences based on their analysis of blood types in Ireland.[8]

2.5.2   The Ape-Human Split

The Irish analyses provide an example of reconstructing population history from anthropometric and genetic data. Many such studies have been conducted by anthropologists to learn more about the impact of historical events on local and regional patterns of genetic variation. Some examples include analyses of the spread of populations and genes into Europe during the development and spread of agriculture, gene flow between European settlers and enslaved Africans in early American history, and the origin of Native Americans, among others.

The genetic history of different species is also a common focus of genetic studies. Of particular interest to anthropologists are studies comparing the genetics of humans to other primate species, in an attempt to reconstruct the history of speciation among the higher primates. One major finding concerns the relationship of humans to the great apes—the Asian great ape, the orangutan, and the three species of African great apes, the gorilla, chimpanzee, and bonobo. All of these species (as well as the lesser ape, the gibbon) are classified as hominoids, primates that have certain characteristics in common. Hominoids lack tails and have mobile shoulder anatomy that allows them to raise their arms above their heads to climb and hang from below. The fact that humans are hominoids is clear at any playground; children frequently hang from their arms on the so-called "monkey bars" in the same manner that apes hang from branches (because monkeys tend to walk on top of

Figure 2.11. *Genetic distance map of mitochondrial DNA differences between humans and the living great apes. The distances were derived from a 684-bp mtDNA sequence reported by Ruvolo et al. (1993) for six humans, three chimpanzees, three bonobos, two gorillas, and one orangutan. The distance matrix was initially computed between all 15 sequences but intraspecies differences are not visible in the graph.*

branches rather than hang below them, this piece of playground equipment ought to be called "hominoid bars").

Within the hominoids, humans have traditionally been considered apart from the great apes. Some of the reasons are obvious when considering overall physical appearance—humans walk on two limbs, are less hairy, and have large brains and small canine teeth, whereas the great apes walk on four limbs, are hairier, and have smaller brains and larger canine teeth. From the perspective of a classification based on overall physical similarity, the division of humans and great apes makes sense, but such classifications do not always reflect evolutionary history. We need to filter out traits that reflect common ancestry versus those that reflect primitive retentions or parallel evolutionary changes. Superficially, humans look quite distinct from the great apes as a group, but closer anatomical comparisons suggested to some scientists, such as Charles Darwin and Thomas Henry Huxley, that humans and the African apes were actually more closely related than either was to the Asian orangutan.

There are now detailed methods for looking at evolutionary relationships among living species based on genetic comparisons. The genetics of humans and the African apes are more similar to each other than either is to the genetics of the orangutan. The close relationship between humans and African apes (particularly the chimpanzee) was shown first using blood proteins and then later confirmed with DNA analysis. An example of such an analysis is presented in Fig. 2.11, which presents a multidimensional scaling plot (a method similar to principal coordinates analysis) of the number of nucleotide differences in a 684-bp sequence of mitochondrial DNA, based on data reported by Maryellen Ruvolo and colleagues.[9] The clustering of humans with the African apes and apart from the orangutan is immediately obvious.

What evolutionary interpretation can be made regarding the genetic similarity of humans to the African apes? Quite simply, closer genetic similarity reflects more recent common ancestry, much in the same way that your DNA is more similar to that of a sibling than that of a first cousin—you and your sibling have more recent common ancestry (your parents) than you and your first cousin (one set of grandparents). On average, the greater the genetic similarity between two species, the more closely related they are, and the more recently they share a common ancestor. Genetic evidence shows clearly that humans and African apes are more similar genetically to each other than either is to the orangutan. This means that humans and African apes share a more recent common ancestor with each other and a more distant common ancestor with the orangutan. Despite the overall physical similarity of the great apes, they do not form a group separate from humans, because the African apes and humans are the most similar to each other.

The greater similarity of humans and African apes had been shown in the early 1960s by the work of Morris Goodman, who worked on blood proteins. Additional work confirmed this observation, including DNA studies that demonstrated that humans and chimpanzees share over 98% of their genes. The close kinship of humans and African apes took on greater significance with the work of Vincent Sarich and Allan Wilson in the mid 1960s. Genetic similarity provided a clue as to the evolutionary history of living species. It became clear that the genetic data supported a model in which a common ancestor of hominoids split into two lines, one leading to the modern-day orangutan and the other leading to the common ancestor of African apes and humans, who split from each other more recently in time. Sarich and Wilson took the analysis one step further by showing that for certain molecular markers the estimated rates of evolutionary change were more or less constant. This constancy allowed them to estimate the approximate dates at which species split from one another from the molecular data.[10]

The basic idea of such a "molecular clock" is that the genetic data provide us with a relative idea of the timing of speciation, which can then be calibrated using the fossil record. For example, imagine that we have genetic distances between three living species—A, B, and C. Further suppose that we compute the genetic distances between each of the three species, getting the following numbers (which in this case reflect relative genetic differences):

Distance between A and B = 3

Distance between A and C = 30

Distance between B and C = 30

We would first infer that species A and species B share a more common recent ancestor because they are more closely related to each other than either is to species C (with genetic distances, the lower the number, the greater the similarity). We would suggest an evolutionary history in which species C split from a common ancestor first, followed by a later split between species A and B. Expressing this more quantitatively, we can say that the distance between A and B is one-tenth that between either A and C or between B and C. Now, suppose that we knew from the fossil evidence that the lines leading to species C and the common ancestor of species A and B split 30 million years ago. Because the distance between A and B is one-tenth that between either and species C, we can estimate that species A and B split from one another $30 \times 1/10 = 3$ million years ago. If we have a reconstructed evolutionary tree, and if we have evidence of constancy in rates, then we can use a single calibration point from the fossil record to fill in the dates for the rest of the tree.

Sarich and Wilson did this using species differences in the albumin protein, and they concluded that the ape and hominid lines split roughly four to six million years ago (hominids refer to humans and human ancestors that lived after the split from the African ape line). This date was initially very controversial because many paleoanthropologists were then suggesting that hominids split off 15–20 million years ago, based in large part on fossil remains assigned to the genus "*Ramapithecus*" (later reclassified into the genus *Sivapithecus*). "*Ramapithecus*" was viewed by some (not all) anthropologists as a very early hominid dating back at least 14 million years ago. The suggestion by Sarich and Wilson that hominids diverged much more recently was not accepted by many paleoanthropologists; clearly, something was wrong, either the fossils or the molecular dating. A debate soon followed, perhaps made more acrimonious by Sarich's statement that "the body of molecular evidence on the *Homo-Pan* relationship is sufficiently extensive so that one no longer has the option of considering a fossil specimen older than about eight million years a hominid *no matter what it looks like* [italics in original]."[11] This statement was taken by many paleoanthropologists as a dismissal of their views and evidence.

As it turns out, Sarich and Wilson were correct, and continued research confirms the divergence of hominids about five to seven million years ago. The debate was not settled exclusively by molecular evidence; continued discovery and analysis of fossils also confirmed a recent separation. "*Ramapithecus*" is now known to be one of many fossil apes that lived before the ape-human split. In this case, the genetic and fossil evidence converged to a common solution. However, the arguments over which evidence is best, fossil or genetic, continue to occur in the context of the modern human origins debate.

2.5.3   Different Approaches to Population History

These two examples both illustrate briefly how genetic data can be used to make inferences regarding population history. Both examples involve comparison of genetic distances between different units of analysis, different populations within a species for the Irish example and different species for the ape-human comparison example. In addition, both examples rely on the principle that closer genetic similarity reflects shared ancestry.

The two examples are different, however, in their interpretation of the underlying mode of evolution. In the Irish example, anthropometric differences between populations were interpreted as a reflection of patterns of gene flow accompanying the movement of people into Ireland. In the ape-human example, genetic differences reflect the accumulation of mutation and the effect of genetic drift in several separate species. These two examples reflect two different approaches to the analysis of population history, one suited for comparing populations *within* a species, and the other used to compare variation *between* different species. Elsewhere, I have referred to these approaches as the "population structure" and "phylogenetic branching" perspectives.[12] The population structure perspective views genetic differences among populations within a species as reflecting the evolution of populations connected by gene flow. Genetic differences (for neutral traits) reflect the balance of mutation, genetic drift, and gene flow. The phylogenetic branching perspective views genetic differences as having arisen through the process of speciation. Here, populations become reproductively isolated and mutational differences accumulate over time. The process of evolution is a series of bifurcating splits, with parent species giving rise to new daughter species over time.

It is clear that we would not switch these approaches for the two examples discussed here. We would not consider that the separation of the Irish midlands in Fig. 2.9 reflects a branching process, whereby the midlands split off as a separate evolutionary entity from the remainder of Ireland, because we know from historical and current observations that all human populations are interconnected via gene flow. Applying the phylogenetic branching model would impose a "tree" interpretation that would be completely inaccurate. Likewise, we would not want to interpret the DNA differences in Fig. 2.11 under a population structure perspective, suggesting that African apes and humans have greater gene flow between them than to orangutans, because we know that they are separate species and do not experience gene flow between them.

The choice of an appropriate model, and related methods of analysis, would seem crystal clear. Are the units of analysis different populations within a species or different species? If we wish to investigate the genetics of humans living in a group of villages on a Pacific island, the population structure perspective is the obvious choice. On the other hand, if we are interested in comparing the genetics of different species of monkeys, then the phylogenetic branching perspective is the appropriate model. The issue is less clear, however, when considering the modern human origins debate, because we don't know whether various archaic human populations constituted different species or not. Indeed, that is one of the basic questions we seek to answer. If we choose one particular interpretive model over the other, then we are imposing the answer on the question! A common theme throughout this book is the need to examine alternative models when considering the genetic evidence on modern human origins. The fact that the data are compatible with one model does not necessarily mean that we have chosen the correct model. We can only use compatibility with a model as proof of that model if alternative models are not compatible.

2.6 SUMMARY

The relatively simple four-letter "alphabet" of DNA is the basis for the genetic structure of life. Genetic diversity can be examined on a variety of levels ranging from the actual DNA sequences to less direct, but still informative, measures of genotype and phenotype. At a micro-level, evolution consists of changes in the relative frequency of different alleles; such change is due to the balance between the four evolutionary forces: mutation, natural selection, genetic drift, and gene flow. The interaction of the evolutionary forces can lead to changes within species over time, and also to the "birth" of a new species though reproductive isolation and genetic divergence.

Evolutionary history can be reconstructed by examining genetic variation in living populations as reflections of the past. Different approaches are used depending on whether we are looking at microevolutionary changes within a single species (the Irish example) or are looking at macroevolutionary changes that led to the formation of separate species (the ape-human split example). The current debate over the origin of modern humans is a bit more complicated because we don't know a priori which model is the most appropriate. Are modern humans the result of a speciation event 200,000 years ago, or do our roots extend further back in time in a single species spread across the Old World for the past two million years?

Chapter *3*

The Modern Human Origins Debate

Journalism students are advised to focus their stories on six things—who, what, where, when, why, and how. This is also good advice when considering the modern human origins debate. Of the fossils discovered to date, who are modern humans and who are other forms (and what is the relationship between them)? What exactly does it mean to be a modern human? Where and when did modern humans arise? These are all excellent questions, and it is often frustrating to some of my students that they cannot always be answered clearly. Indeed, much of the debate over modern human origins concerns the basic issues of who, what, where, and when? Answers to these questions affect our answer to the overriding question of this book—*how* modern humans evolved.

To start answering such questions, we have to deal with the more basic question of what humans are, modern or otherwise. A frequent question from my students deals with a very basic question, one they expect I will be able to answer quickly and definitely: How long have humans been around? This is not an easy question to answer because it requires that we be specific in our definition of "human." How are humans defined? A typical list includes both biological (bipedalism, large brains, small canine teeth) and behavioral (language, tool making) attributes. Some characteristics are difficult to define exactly because they are found to some extent in other species (for example, apes also make and use simple tools). There is also the problem of quantitative specificity; for example, how large a brain qualifies as human? Does overall size matter, or is size relative to body size a more appropriate measure?

Apart from such definitional problems, the question is difficult because it has no single answer. Our ancestors began walking upright at least 4.2–4.4 million years ago. If we define "human" solely in terms of bipedalism, I could tell my students that humans have been around over four million years. However, these early hominids were similar to us in some ways but not in others. Over four million years ago, our ancestors still had relatively small brains and large canine teeth, which are considered apelike characteristics. How about if we define "human" in terms of a significant increase in brain size and the development of stone tool technology? If so, then the answer is 2.0–2.5 million years ago. On the other hand, if we use small and nonprojecting canines, the answer is roughly three million years ago. If we take other attributes, such as symbolic expression through art, the answer is again different (roughly 40,000 years). Depending on our definition, we will wind up with different dates. This is expected, because human evolution did *not* involve an instantaneous change from ape to human at a single point in time. Different traits appeared at different times, reflecting the mosaic nature of evolution.

The questions of modern human origins must be viewed in the broader context of hominid evolution.[1] Although evolutionary transitions can be relatively quick (in geological time), they are not instantaneous, and the origin of modern humans must consider what came before. The first part of this chapter briefly reviews what we know about the early parts of hominid evolution, followed by a more detailed look at evolution within the genus *Homo* over the past two million years. It is within this context that we can address the questions of what modern humans are, where and when they first appeared, and how they evolved.

3.1 GEOLOGIC TIME

An illuminating experience for many of my students comes when they take a close look at the enormity of geologic time and the very small fraction of time that hominids have been present.[2] Figure 3.1 shows a standard time scale, illustrating how geologists divide Earth's history into two eons, the Precambrian and the Phanerozoic. Vertebrate origins and evolution took place within the Phanerozoic Eon, which began 545 mya ("mya" = "millions of years ago"). Note that the last 545 million years accounts only for 12% of earth's history. The Phanerozoic Eon is

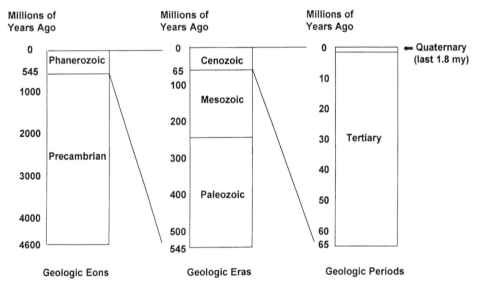

Figure 3.1. Geologic time scale for Earth's history (Source: Schopf 1992).

further divided into three geologic eras, the Paleozoic, Mesozoic, and Cenozoic. Major events in the Paleozoic Era include the diversification of early vertebrates and the origin of primitive fish, amphibians, and reptiles. The Mesozoic Era (the "Age of Dinosaurs") saw the diversification of many reptiles, including dinosaurs, as well as the origin of mammals. The Cenozoic Era (the "Age of Mammals") covers only the last 65 million years, or 1.4% of earth's history. The Cenozoic Era is further divided into two geologic periods, the Tertiary and the Quaternary, each of which is divided further into several geologic epochs.

Our earliest evidence of hominids (4.2–4.4 mya) is in the Pliocene epoch (1.8–5.0 mya) of the Tertiary period. This means that hominids have only been around 0.09% of Earth's history. The first evidence of the genus *Homo*, characterized by an increase in brain size and the development of stone tool technology, is dated to 2.5–2.0 mya—roughly 0.05% of earth's history. The proportion of time dealing with the modern human origins debate is roughly 200,000 years, which seems like a very long time in terms of short life spans but is actually only 0.004% of Earth's history!

3.2 EARLY HOMINID EVOLUTION

The first hominids were bipedal, although they probably spent considerable time still climbing in trees for food and safety. These hominids had certain anatomic characteristics typical of human bipedalism, including a wider pelvis and upper legs that angled out from the knees. In other respects, they were still very apelike, having small brains and large, protruding faces with larger canines than found in later hominids. All early hominids, classified into the genera *Ardipithecus* and *Australopithecus*, have been found in Africa, the homeland of hominid evolution. It is not until much later that hominids began dispersing out of Africa. The oldest suggested hominid, *Ardipithecus ramidus*, lived in east Africa and dates back 4.4 million years. Although fragmentary, the *Ardipithecus* fossils show very primitive, apelike teeth and the suggestion of bipedalism. More definitive evidence of bipedalism is found 4.2 mya with the species *Australopithecus anamensis*. *A. anamensis* is the oldest currently known australopithecine, a group of hominids characterized by bipedalism, small brains, and large faces (relative to later hominids). The term "australopithecine" (and the genus name *Australopithecus*) translates as "southern ape," a name given because of the apelike traits and the fact that the first australopithecine ever discovered was found in South Africa.

By three million years ago, *A. anamensis* appears to have evolved into the species *Australopithecus afarensis*, best known by the famous fossil "Lucy," a 40% complete skeleton (Fig. 3.2). *A. afarensis* had somewhat less apelike teeth than *A. anamensis*. Between 3.0 and 2.5 mya, there was diversification of the australopithecines, leading to a number of different species whose relationships are still being debated. Some of these species are commonly termed "robust" australopithecines after their massive back teeth, jaws, and chewing muscles, apparently adaptations for processing hard-to-chew foods. The robust australopithecines appear about 2.5 mya and eventually become extinct roughly 1 mya. Another australopithecine species, *Australopithecus africanus* (Fig. 3.3) is not as robust, having a more gracile skull and dental and facial proportions similar to later hominids classified in the genus *Homo*. A number of paleoanthropologists have long suggested that *A. africanus* is a likely

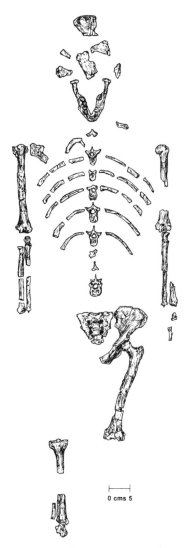

Figure 3.2. Australopithecus afarensis *specimen A.L.-288-1 from Hadar, Ethiopia, more commonly known as "Lucy." This specimen dates to roughly three million years ago. (from Larsen CS, Matter RM, and Gebo DL, Human Origins: The Fossil Remains, third edition, p. 50. Copyright © 1998, 1991, 1985 by Waveland Press, Inc., Prospect Heights, IL. All rights reserved.)*

direct ancestor to *Homo*, although a newly discovered species (*Australopithecus garhi*) is another possible candidate.

Although definitely hominids, can the australopithecines be called "human?" The answer to this question depends on the specific definitions that are used. The major human characteristics appearing during australopithecine evolution are bipedalism and the reduction in the size of the canine teeth. There is no definite association of australopithecines with early stone tools, although some paleoanthropologists have suggested that some of them did have the manual dexterity required for stone tool

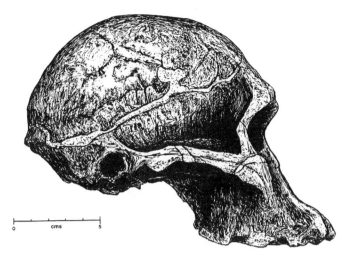

Figure 3.3. Australopithecus africanus *specimen STS 5 from Sterkfontein, Republic of South Africa. This specimen dates to approximately 2.3–2.8 million years ago. (from Larsen CS, Matter RM, and Gebo DL, Human Origins: The Fossil Remains, third edition, p. 60 Copyright © 1998, 1991, 1985 by Waveland Press, Inc., Prospect Heights, IL. All rights reserved.)*

manufacture. Studies of brain anatomy (made possible by the impressions left on the inside of skulls) are controversial but suggest that their brains were still very apelike in structure. There is no evidence of organized hunting, use of fire, or symbolic expression. On the basis of the absence of these characteristics (and acknowledging the arbitrary nature of definitions of "human"), I would argue that the australopithecines, although hominid, are not "human," but rather a group ancestral to later hominids whose biology and behavior link characterize them as human. This later group, having larger brains, is classified in the genus *Homo*.

3.3 EVOLUTION OF THE GENUS *HOMO*

The genus *Homo* has an African origin. All of the early fossils assigned to *Homo* have been found in Africa, followed by later fossils in Asia and Europe. The fossil record shows that hominid evolution begins in Africa with the australopithecines. By 2.5 mya, some species of australopithecine (*A. africanus*? *A. garhi*?) evolves into the first members of the genus *Homo*. Evidence for the initial origin of *Homo* is fragmentary, but we do have some fossils assigned in a general sense to *Homo* in Africa dating back close to 2.5 mya. Although these fossils provide evidence for the initial appearance of *Homo*, they are not complete enough to assign to a given species. It is also at this time that we see the first evidence of simple manufactured stone tools.

By 2 mya, we have evidence of three different species of *Homo* living in East Africa. One species, *Homo habilis*, has a large brain than earlier hominids but retains certain primitive traits in their skeletons (e.g., relatively long arms, an apelike characteristic also found in the australopithecines). Another species, *Homo rudolfensis*, has an even larger brain, but with australopithecine-like dental and facial proportions. One of these two species may represent the continued survival of the first *Homo* species, but, as noted above, the earliest remains are too fragmentary to determine which (if either) species is a link between australopithecines and the third species to consider, *Homo erectus*, who is our direct ancestor.[3]

3.3.1 Early Humans—*Homo erectus*

Homo erectus has a noticeably larger braincase and smaller face than the australopithecines although still different from living humans (Fig. 3.4). The average cranial volume of *H. erectus* is 1000 cm³, roughly three-fourths the average size of living humans (1350 cm³), although individual specimens show considerable variation, ranging from about 750 cm³ to 1250 cm³, and overlapping the range of living humans (roughly 1000–2000 cm³). From the neck down, *H. erectus* has long legs and body proportions similar to those of living humans. These changes suggest that *H. erectus* was the first hominid species to be fully committed to bipedalism, particularly long-distance striding, which may be related to the fact that *H. erectus* is the first hominid species to leave Africa.[4]

It has long been known that some populations of *H. erectus* moved into Asia during the past million years. In 1994, however, new analysis suggested that *H. erectus* might have moved into Indonesia between 1.6 and 1.8 mya.[5] There is also a suggestion of early movement into China, perhaps by 1.1 to 1.9 mya.[6] The issue of European *H. erectus* was controversial; fossils dating back to 500,000 and 800,000 years ago are known from several European sites, although it was not always clear whether these fossils should be classified as *Homo erectus* or *Homo sapiens*. Very early dates of 1.7 mya were found for *Homo* fossils from eastern fringes of Europe, although these remains could not be accurately classified. This situation changed in May 2000, when Leo Gabunia and colleagues reported the discovery of two *H. erectus* skulls at the site of Dmanisi in the Republic of Georgia dating to 1.7 mya.[7] Both skulls are very similar to fossils of similar age from East Africa. Collectively, these studies show that *H. erectus* was on the move shortly after its initial appear-

(a)   (b)

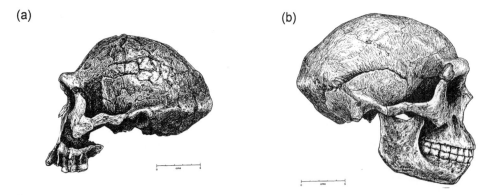

Figure 3.4. Homo erectus. (a) KNM-ER 3733 from East Turkana, Kenya, dating to 1.8 million years. (b) Skull reconstruction from Zhoukoudian, People's Republic of China, dating to approximately 300,000–500,000 years. (from Larsen CS, Matter RM, and Gebo DL, Human Origins: The Fossil Remains, third edition, p. 93 and p. 110. Copyright © 1998, 1991, 1985 by Waveland Press, Inc., Prospect Heights, IL. All rights reserved.)

ance in Africa, testimony to a species capable of long-distance movement and adaptation to a variety of environments across much of the Old World.

There is some debate about the differences between African and Asian forms of *H. erectus*. Some paleoanthropologists argue that the geographic differences reflect the existence of two separate species, a species named *Homo ergaster* in Africa and *Homo erectus* in Asia.[8] According to this view, *H. ergaster* led to two different lineages, one ancestral to later humans and the other leading to *H. erectus*, which was an evolutionary side branch in human evolution. Others argue that the geographic differences are minor and of the level expected in a species spread out over two continents.[9] The recent discovery of the Dmanisi skulls fits with this idea; they are more similar to African *H. erectus* than to Asian *H. erectus* but are also closer in both time and space to the African specimens, suggesting a pattern of isolation by distance—the further two populations are apart from one another, the less similar they will be because of the limiting effect of geographic distance on migration. As we shall see, the issue of the number of species in the genus *Homo* is central to the modern human origins debate. For the moment, I will use the label *Homo erectus* to refer to all of these early human fossils.

Archaeological evidence has helped flesh out some aspects of the lifestyle and culture of *H. erectus*. The earliest evidence of stone tool technology consists of simple chopping tools of the Oldowan tradition, formed by the removal of one or more flakes from a stone core. Some archaeologists have suggested that the small flakes were more often used as cutting tools, and the Oldowan choppers are what were left over after removal of the flakes. By 1.4 mya, *H. erectus* had invented a new method of making stone tools, characterized by the removal of stone flakes from both sides of the core (bifacial flaking). Known as Acheulian tools, these tools were sharp and symmetric in shape. One abundant type of Acheulian tool is the somewhat misnamed "hand axe," a large tool used for a variety of cutting purposes. Acheulian tools, including hand axes, are found later in time as well with *Homo sapiens*; there is not a direct correlation between tool type and species name. For many years, the geographic distribution of bifacial tools puzzled anthropologists. Acheulian tools were found in Africa and Europe but not in East Asia, where instead simpler chopping tools were found. A variety of ideas have been proposed to explain this geographic difference, ranging from different species with different behavioral capabilities to differences in the type of raw material found in different locations. Recent discoveries have suggested that the West-East difference may not have been as great as we once thought. Hou Yamei and colleagues reported in March 2000 the discovery of Acheulian-like stone tools dating back 800,000 years ago in China.[10] Once again, we should be reminded that earlier ideas can be rejected by the accumulation of new data, and we are far from having a complete fossil or archaeological record.

Many anthropologists agree that *H. erectus* practiced hunting, although there is still debate over their capability. Evidence of hunting comes from Africa 1.5 mya, where there is a noticeable change in the nature of butchered animal bones. Before this time, many of the butchered bones were those frequently left over from carnivore kills, suggesting that *Homo* was scavenging. Although scavenging undoubtedly continued as an adaptive strategy, later animals have more bones represented, suggesting more complete utilization of a carcass. The debate over the relative importance of scavenging versus hunting will continue, but either way the finding of butchered animal bones with characteristic stone tool cut marks shows that *H. erectus* had added animal protein to its diet.

3.3.2 "Archaic" Humans

Fossils assigned to *Homo erectus* date back close to two million years. Most paleoanthropologists see *H. erectus* (in the broad sense including both African and Asian forms) as ancestral to *Homo sapiens*. The nature of this relationship is very controversial, with some arguing for a gradual change from one form into another and others arguing that there was one or more distinct speciation events. As such, there is often disagreement about what species a particular fossil should be assigned to. This problem means that although there is agreement over the timing of the origin of *H. erectus*, there is less consensus over its total time range, because under a model of anagenesis the cutoff is somewhat arbitrary, whereas under cladogenesis a species has a definite beginning and a definite ending.

A good example of this problem is the fossil crania from Ngandong, Indonesia. These crania (which lack faces) are similar in some ways to earlier specimens of Asian *H. erectus* but also have greater cranial expansion more typical of *H. sapiens*. Some classify these remains as late *H. erectus* and others as archaic *H. sapiens*. Complicating the analysis are new geologic dating estimates that suggest that they may be as recent as 27,000–53,000 years old rather than earlier estimates of roughly 200,000 years old.[11] To some, the late appearance of fossils resembling *H. erectus* is suggestive of replacement by modern humans in Southeast Asia within the past 100,000 years.

After the emergence of *H. erectus*, there was little change in average brain size until about 700,000 years ago, at which point there was a rapid increase (Fig. 3.5).[12] By roughly 400,000–500,000 years ago, we begin to see evidence of fossil hominids that are different in some ways from *H. erectus*, with average cranial capacities within the range of living humans. There is considerable disagreement over what to call these fossils because they do not fit neatly into either *Homo erectus* (as based on earlier fossils) or *Homo sapiens* (as based on living humans). Indeed, many individual fossils have been referred to over the years as "late *Homo erectus*" or "early *Homo sapiens*." It has become common to refer to these hominids as "archaic" humans or, in some cases, "archaic *Homo sapiens*." This term captures the fact that in some ways these fossils are very similar to living humans (such as a large brain), but in other ways they are different.

Archaic humans generally had large brain sizes similar to modern humans but a rather different cranial shape. Archaics often have a low cranial vault, large brow ridges, a sloping forehead, and a robust face (Fig. 3.6). One well-known group of archaic humans was the Neandertals (Figure 3.6c),[13] who lived in Europe and the Middle East from roughly 150,000 to 28,000 years ago.[14] They shared certain characteristics, such as a low cranial vault and smaller brow ridges, with other archaic groups and additionally had large noses, midfacial projection, and often a bulging region at the back of the skull (known as an "occipital bun"). Neandertal skeletons tend to be rather short and stocky, and they were generally rather muscular. Neandertals made and used sophisticated stone tools known as Mousterian tools; these tools were made by the careful preparation of a stone core from which multiple tools could be flaked (an early form of mass production!). Neandertals buried their

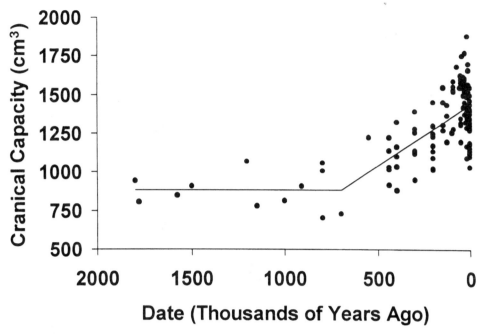

Figure 3.5. *Evolution of cranial capacity (cubic centimeters) in the genus Homo over the past two million years. This plot is based on 136 specimens ranging in age from 1000 years ago to 1.8 million years ago. Source of data: Ruff et al. (1997). The solid line is the best fit of a nonlinear piecewise regression line. See chapter notes for computational details.*

dead, an early indication of some sort of symbolic behavior even though we do not know their exact motivations.

3.3.3 "Modern" Humans

Anatomically modern humans are often defined by a high, well-rounded skull, vertical forehead, reduced face, teeth, and brow ridges, and a distinct chin. These are general characteristics, and there are a number of specimens that show variation. In addition, there are living humans who do not always fit this typological image of "anatomically modern," leading some paleoanthropologists, such as Milford Wolpoff and Rachel Caspari, to question whether we can even use such labels as "archaic" and "modern" when dealing with variation across time and space.[15] Figure 3.7 shows three skulls widely considered modern. The first (Fig. 3.7a) is Skhul 5, a skull from Israel dating back 92,000 years. Although this skull shows a high cranial vault, a vertical forehead, and a chin, it also has a relatively large face and brow ridges. This specimen is considered by some to be an early modern, still possessing some archaic features. The second skull (Fig. 3.7b) is from the French site of Cro-Magnon, dating between 23,000 and 27,000 years ago. Often portrayed as a typical modern human, this skull is high and well rounded, with small brow ridges, a vertical forehead, and a noticeable chin. The third skull (Fig. 3.7c) is from the Kow Swamp site in Australia, dating to between 9000 and 13,000 years ago. This specimen has a sloping forehead, a larger face, and large brow ridges. Obviously, the

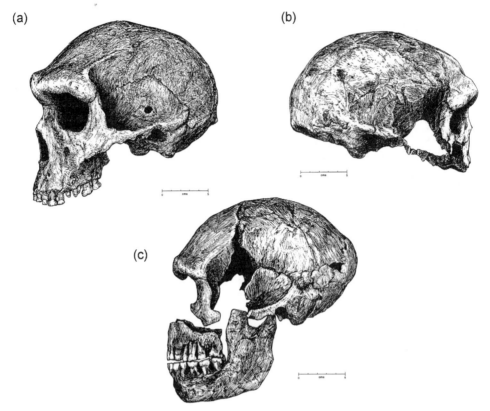

Figure 3.6. *Archaic humans. (a) Broken Hill 1 from Kabwe, Zambia, dating to roughly 125,000 years ago or older. (b) Dali from Shaanxi Province, People's Republic of China, dating to roughly 200,000 years ago. (c) La Quina 5 from Charente, France, dating to roughly 40,000–55,000 years ago. This specimen is a Neandertal. (from Larsen CS, Matter RM, and Gebo DL, Human Origins: The Fossil Remains, third edition, p. 111, p. 128, and p. 147. Copyright © 1998, 1991, 1985 by Waveland Press, Inc., Prospect Heights, IL. All rights reserved.)*

differences between archaic and modern humans are not always clear, and we must keep in mind variation that exists across both time and space.

Despite a blurred boundary between archaic and modern humans, many anthropologists agree that the first indications of more modernlike anatomy appear first in Africa, perhaps as early as 130,000 years ago. The next oldest moderns have been found in the Middle East, dating back over 90,000 years ago. Modern humans remains in Europe date to between 30,000 and 40,000 years. On the surface, the geographic distribution of modern humans shows a striking pattern of an early emergence in Africa, followed by movement through the Middle East, and then later into other parts of the Old World. There are disagreements, however, with this simple pattern, including arguments about the supposed archaic or modern attributes of various fossils, as well as the accuracy of the dates.[16] Furthermore, we must guard against the possibility that our tendency to classify and pigeonhole fossils may be giving us a distorted image of a transition from archaic to modern. Are the two morphologies that different, and do they change that abruptly in different parts of the world? Are we imposing a dichotomy on a continuous process, much the same way

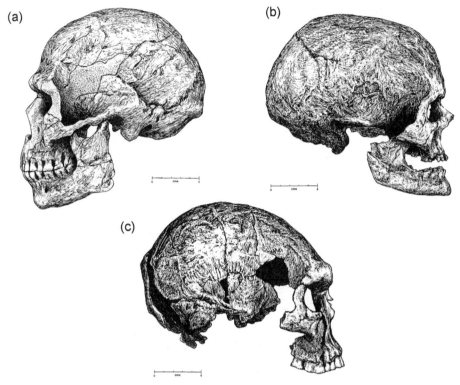

Figure 3.7. *Anatomically modern humans. (a) Skhul 5 from Wadi el-Mughara, Israel, dating to 90,000 years ago. (b) Cro-Magnon 1 from Dordogne, France, dating to 23,000 to 27,000 years ago (the jaw looks odd because of missing teeth and old age). (c) Kow Swamp 1 from Victoria, Australia, dating to 9,000 to 13,000 years ago. (from Larsen CS, Matter RM, and Gebo DL, Human Origins: The Fossil Remains, third edition, p. 157, p. 171, and p. 180. Copyright © 1998, 1991, 1985 by Waveland Press, Inc., Prospect Heights, IL. All rights reserved.)*

we might classify people as short, medium, and tall, giving the illusion of three types from a continuous distribution?

The archaeology of modern humans is equally problematic. The term Lower Paleolithic is often used to describe the tool industries of *Homo erectus*, whereas Middle Paleolithic is used as a label for the tool industries used by archaic and early modern humans. The Upper Paleolithic is associated with a variety of new technologies and behaviors, including finely made blade tools, bone tools, and cave art. Although some anthropologists argue that the Upper Paleolithic is associated exclusively with anatomically modern humans, others note evidence of an association with some archaic samples. Again, definitional problems and the range of variation make consensus difficult. There is also the problem of defining and identifying the change from Middle Paleolithic to Upper Paleolithic; some see this transition as abrupt, whereas others see it as a more gradual transition.

3.3.4 Lumpers and Splitters

The preceding review of the human evolutionary record is meant only as a summary; most of the fine details and arguments have been glossed over in the interest of brevity, because a complete review would (and has!) taken volumes. The main point is that we have a general consensus of some major evolutionary trends (such as an increase in brain size, a shift in the shape of the skull and the reduction in robustness, and increasing technological and symbolic behaviors) but disagreement over the fine details of the picture. For the purpose of this book, the primary debate is over the number of species of *Homo* represented over the past two million years (excluding the australopithecine-like species *Homo habilis* and *Homo rudolfensis*).

The debate over the number of species in the genus *Homo* is often characterized by a debate between "lumpers" and "splitters" and their respective interpretations of biological variation.[17] Lumpers tend to perceive variation as a normal consequence of the evolutionary process and are hesitant to assign different species names to what might actually represent variability within a single, evolving species. Lumpers also tend to view macroevolution as a gradual process, with transitions from one form to another over time. Splitters tend to perceive variation as a signature of cladogenesis and view macroevolution as a series of successive speciations. Splitters also tend to advocate the model of macroevolution known as punctuated equilibrium, which associates most evolutionary change with a relatively rapid speciation, followed by a period of relative stasis until the next speciation (or extinction). The debate between lumpers and splitters also concerns the debate over the very nature and definition of species. Should species names be reserved exclusively for a reproductively separate lineage, or can species names be used as rough labels for overall morphology?

If we confine our use of species names to separate evolutionary lineages, then how can we determine whether different fossil samples actually represent different species? The biological species concept involves reproductive isolation—if two populations are capable of mating naturally and producing fertile offspring, then they belong to the same species. This cannot be determined directly from the fossil record, as there is no way of telling whether a given Neandertal sample was interfertile with a given contemporary sample of modern humans. Instead, we must rely on evidence on biological variation in living species and use it as a yardstick with which to compare variation in the fossil record. Although a variety of statistical approaches exist to answer such questions, it is not always clear *what* level of variation is sufficient to classify fossils into different species. Ian Tattersall argues that because many living primate species are physically similar to each other, we would tend to incorrectly lump them together if we found them in the fossil record. Consequently, we will always tend to underestimate the number of past species. Others, including Milford Wolpoff, argue that the biological species concept, developed to classify living species, cannot easily be applied to fossils.

Given differences in definitions, methods, and underlying philosophy, it is no surprise that different scholars partition the past two million years of evolution in the genus *Homo* in different ways, as illustrated in Fig. 3.8. The model on the left (Fig. 3.8a) represents the ultimate in lumping; according to this view, there has been only one evolutionary lineage of *Homo* over the past two million years. This species arose first in Africa and then spread across the Old World, differentiating and adapting to local environments, but remained part of a single widespread species whose populations were interconnected by gene flow. Because there was only one species at any given point in time, species names such as *Homo erectus* and *Homo sapiens*

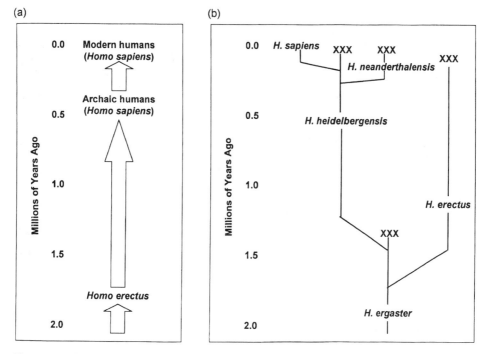

Figure 3.8. *Alternative models of evolution in the genus* Homo. *(a) A "lumpers" perspective, in which all* Homo *fossils are part of a single evolving lineage. There is only one species at any point in time. Names such as* Homo erectus, *archaic humans, and modern humans are simply taken as labels for general stages in the evolution of* Homo, *and the division between them is somewhat arbitrary. (b) A "splitters" perspective, in which variation in* Homo *fossils represents different species, such that two or more species lived at the same time. According to this view, modern humans* (Homo sapiens) *arose recently as a new species branching off of the ancestral species* Homo heidelbergensis.

are meant only as labels to represent different stages along a continuous evolutionary line. In this context, species names are meant as labels of morphology and not to indicate separate reproductive populations living at the same time. In some forms of this model, the entire *lineage* is given the name *Homo sapiens*, although many paleoanthropologists prefer to keep the label *Homo erectus* to refer to the early members of the lineage.[18]

Figure 3.8b represents a completely different view, this time from a splitter's perspective. According to this view, there have been two (or more) hominid species during most of the past two million years, and as we accumulate more fossils, we will find even more. The evolution of *Homo* was characterized by cladogenesis, with new species being "born" and other species dying off. Here, the first species, *Homo ergaster*, lived in Africa, and an Asian species, *Homo erectus*, branched off, ultimately becoming extinct within the past 100,000 years or so. *Homo ergaster* ultimately became extinct, but not before giving rise to a new species, *Homo heidelbergensis*, that includes archaic humans throughout parts of Africa, Europe, and possibly Asia. This species later gave rise to two other species, *Homo neanderthalensis* (the Neandertals) in Europe, and *Homo sapiens*, probably in Africa. Today, only we (*Homo sapiens*) remain—all other species have died out. Four different species lived within the past 200,000 years: *H. erectus*, *H. heidelbergensis*, *H. neanderthalensis*, and *H. sapiens*. This view is not even the most extreme view of splitters; some suggest that another species, *Homo antecessor*, lies between *H. ergaster* and *H. heidelbergensis*. Of course, there are many other variants. Some splitters would still lump *H. erectus* and *H. ergaster*, and some lumpers might allow for the possibility that the Neandertals were a separate species.

Given these differing views, the use of terms such as "archaic humans" and "modern humans" varies accordingly. To some, archaics and moderns represent stages in the evolution of a single species over time. To others, the archaics represent several different species and moderns refer only to the latest species, *Homo sapiens*. If we ignore the species names for the moment, we can see that there is some general agreement that archaic humans evolved into modern humans but that there is debate over *which* archaics and *how* this evolution occurred. Did modern humans evolve from a *single* archaic population in Africa through speciation, or did they evolve via anagenesis in more than one part of the Old World? This question forms the crux of the modern human origins debate.

3.4 MODELS OF MODERN HUMAN ORIGINS

Some parts of the modern human origins debate date back to the nineteenth century, when scientists were trying to figure out the evolutionary relationships of the then newly discovered specimens of Neandertals and living humans. In 1864, William King reviewed the anatomy of the original Neandertal skull, concluding that it was best placed in a separate species from living humans—*Homo neanderthalensis* (in a footnote to his 1864 paper, King further suggests that Neandertals were different enough to possibly warrant a separate genus!).[19] Since that time, the evolutionary status of Neandertals has changed several times.[20] By the early 1960s, more and more anthropologists were accepting Neandertals as a subspecies (*H. sapiens neanderthalensis*) different from living humans (*H. sapiens sapiens*). The primary argument at this time was whether *all* archaic humans went through a "Neanderthal phase" or whether the Neandertals represented the evolution of a specific human group located in Europe and the Middle East. Many considered Neandertals somewhat distinct morphologically but still part of the geographically widespread species of *Homo sapiens*.

To a large extent, the question of modern human origins focused almost exclusively on Europe. The Neandertals were compared and contrasted with somewhat younger European fossils more typically anatomically modern, such as the famous Cro-Magnon site (a site that has taken on such significance that modern humans in Europe from this time period are often referred to popularly as "Cro-Magnons"). The geologic framework was also looked at closely during the 1960s and 1970s. The data at the time suggested that the Neandertals "disappeared" about 35,000 years ago, followed 5000 years later by the Cro-Magnons. There was considerable debate over this 5000-year interval. Was this sufficient time for the Neandertals to involve in situ into modern humans, or was the apparent "replacement" of Neandertals indicative of an actual physical replacement from outside of Europe?

The debate continues even though we now have evidence of later Neandertals existing at the same time as more modern-looking specimens, as well as evidence of a gradual reduction in Neandertal characteristics over time. The main point that emerges from the history of paleoanthropology is that the arguments had for a long time focused exclusively on Europe, perhaps because of a much sparser fossil record elsewhere, as well as a Eurocentric view of human evolution. The focus was on Europe because many people thought that Europe was the main stage in human evolution; today, we realize that, to the contrary, Europe was more on the periphery of human evolution. The questions of the evolutionary status of Neandertals remain but are now part of a more global perspective on modern human origins that takes into consideration the important fossil and archaeological record in Africa and Asia. The finding of modern-looking specimens in Africa earlier than in any other region has underscored the importance of the African continent.

Virtually every textbook today reviews the current status of the modern human origin debate in terms of two basic models—African replacement and multiregional evolution.[21] Briefly, the African replacement model hypothesizes that modern humans arose as a new species in Africa roughly 150,000 to 200,000 years ago, followed by geographic expansion outside of Africa replacing all non-African archaic populations. Under this model, *all* of our ancestors 200,000 years ago lived in Africa. The multiregional evolution model proposes that the transition from archaic to modern humans took place within a single species spread out across the Old World. Here, *some* of our ancestors 200,000 years ago lived in Africa, but others lived elsewhere. There are a number of variants of each general model, as well as models often described as "intermediate."

There appears to be a great deal of confusion over these models, often leading to a number of misconceptions, particularly about model predictions. Paper after paper has been written arguing basic definitions and countering misunderstandings of others.[22] Part of the confusion lies in different uses of the same terms, changing definitions, and, of course, personality conflicts among various participants in the debate. I think that another problem in describing and testing these models is that there are really two basic questions that are often lumped together: 1. What was the *mode* of the transition from archaic to modern human? 2. What was the *location and timing* of the transition? The first question concerns the nature of the evolutionary change(s) leading to the emergence of modern humans. Did modern humans evolve through a speciation event and subsequent replacement of other human populations, or did we evolve through a multiregional process? The second question focuses on *where* and *when* the transition occurred. Did the transition occur only in one place (Africa) at one time, primarily in one place and time, or worldwide over time?

3.4.1 The Mode of the Transition—Replacement or Multiregional Evolution?

The primary debate is over the mode of evolutionary change linking archaic and modern human populations. Was this transition between species or within species? That is, do modern humans represent a species (*H. sapiens*) biologically distinct from archaic humans (*H. heidelbergensis*, *H. neanderthalensis*, and possibly others)? If so, then the change resulted from a speciation event, presumably in Africa. Genetic changes, probably in a relatively small and isolated population, led not only to the emergence of modern morphology but also to sufficient genetic divergence to lead to a new and separate species, as defined by the classic definition of the biological species concept. Reproductive isolation would first be required (elimination of gene flow), followed by subsequent genetic divergence due to mutation, drift, and natural selection. According to replacement advocates, this new species grew in size over time, ultimately expanding out of Africa to move into other regions of the Old World. We know, however, that there were other humans living outside of Africa at this time. By definition, this means that the dispersing population from Africa and the preexisting human groups outside of Africa belonged to different species, and could not produce fertile offspring. This doesn't mean that they didn't interact, or even mate, but that offspring did not result, were sterile, or had reduced fitness. Because only modern humans exist in the world today, this means that all other archaic species became extinct; i.e., *Homo sapiens* out of Africa replaced them.

If the emergence of modern humans resulted from a speciation event in Africa followed by replacement outside of Africa, a logical question is how that replacement occurred. The use of the word "replacement" conjures up images of all-out aggression, with one species wiping the other off the face of the earth. The idea of hominid replacement is an old one and a favorite theme in literature and popular culture. A classic portrayal of this concept is in William Golding's novel *The Inheritors*, which tells the story of a sympathetic Neandertal family before their demise at the hands of the evil invading Cro-Magnons. Such stories fit our popular image of highly aggressive "cave men" fighting tooth and nail to the bitter end, an image reminiscent of the opening scene in Stanley Kubrick's film *2001: A Space Odyssey*. If replacement did occur and was violent, it is surprising that we see no strong evidence of it in the fossil and archaeological records. Where are the broken bones and assorted injuries we would expect with such a violent encounter?

Violence is not the only form of replacement. Another possibility might be disease, brought in by newcomers and wrought on other humans that had no genetic resistance or prior immune experience. The disease hypothesis is hard to defend because epidemics tend to be more common in larger and denser populations than would be typical for hunters and gatherers. Replacement can perhaps best be explained by natural selection operating through the competition of two separate species. If modern humans had even a slight advantage in survival and/or reproduction, then they could easily replace another species in a relatively short period of time. Ezra Zubrow used computer simulation of demographic patterns typical of hunter-gatherers to show that even a one percent advantage in the survival/reproduction of modern humans could result in complete replacement in a millennium.[24] Of course, such studies show only what *could* happen and not what actually *did* happen.

The issue of what caused replacement, although an important question, is not the primary issue. Before we can develop and test hypotheses of what caused the replacement, we must first show that replacement actually happened. If replacement did *not* occur, then the question of what caused it becomes irrelevant. Likewise, a failure to demonstrate what could have caused a replacement does not negate the possibility that it happened. The nature of replacement is an unanswered question contingent upon adequate demonstration that replacement occurred. Finally, I emphasize that the issue of complete replacement requires that we are dealing with two reproductively separate species. If archaics and moderns were *not* separate species, then it is doubtful that replacement would have been complete under any replacement hypothesis. History shows that even in the most extreme and violent cases of culture contact human populations are not *completely* wiped out. They might be conquered and assimilated, and even become extinct in the sense of having a distinct culture, but the genes live on.

Multiregional evolution views the evolutionary change from archaics to moderns as having taken place within a geographically widespread single species [regardless of how this lineage is subdivided by species names, which are used in this context as convenient labels (chronospecies), and not indicative of reproductively isolated evolutionary entities]. The multiregional evolution model was developed by Milford Wolpoff, Wu Xinzhi, and Alan Thorne as a general explanation of the evolutionary process by which a species can change as a whole while still maintaining specific regional characteristics.[25] The concept of regional continuity, the continued expression of traits within a given region over time, is central to multiregional evolution. Wolpoff and colleagues found that comparisons of fossil specimens from the same geographic region across wide intervals of time often show us evidence of continuity. Shovel-shaped incisor teeth, for example, are found in living humans across the world but are more frequent in East Asian populations. The same regional pattern also seems to have existed in the past, suggesting genetic continuity across time within that region. In other words, *some* of the ancestors of living Asians were ancient Asians (but not *all*, or even necessarily *most*).[26] Other examples include the distribution of certain dental traits in Europe, high nasal angles in Europe, sloping foreheads in and near Australia, among others. The fossil record shows some evidence of continuity between past and present populations across the Old World.

The origins of regional characteristics, where certain traits are more frequently found in one geographic region, are not hard to understand evolutionarily. Genetic isolation, combined with genetic drift and natural selection, easily causes geographic variation in the underlying allele frequencies.[27] We see many examples of regional characteristics in living humans, ranging from the more frequent occurrence of shovel-shaped incisors in some Asian populations, to the high frequency of the *O* allele of the ABO blood group in Native Americans, to the high frequency of the *cDe* haplotype of the Rhesus blood group in sub-Saharan African populations. Multiregional evolution proposes that geographic variation across the world began close to two million years ago with the initial dispersion of some populations of *Homo* into Asia and Europe. Because founding groups were very small, there was a lot of opportunity for genetic drift. Over time, these regional characteristics persisted to some extent, so that we find evidence of regional continuity when comparing fossil samples across time.

Should we assume that regional continuity would always reflect population history in terms of ancestors and descendants? Not necessarily, because natural selection could mimic this history. If a trait is strongly affected by natural selection then similarities of populations could also reflect similar adaptive responses. A good example is human skin color. Dark-skinned humans are found among both Central Africans and in Australian aborigines. Does this physical similarity necessarily demonstrate shared ancestry? No, because in this case similarity is a reflection of adaptation to a similar environment. Both Central Africans and Australian aborigines live near the equator. Comparative studies of human skin color clearly show that native populations living at or near the equator have the darkest skin color, and there is a clear gradient in skin color in populations further away from the equator.[28] Demonstrations of regional continuity are most informative when we are dealing with traits that are relatively neutral, ensuring that we are likely to see a reflection of population history and not the confounding effect of natural selection.

Observations of regional continuity have been difficult to explain when considering that the fossil record shows evidence of common evolutionary trends across the entire geographic distribution of human ancestors. The best example of this is the clear increase in brain size starting 700,000 years ago. There is evidence of the *same* trend occurring across the Old World. If, according to multiregionalists, all human fossils belonged to a single evolving species, then how could the same trend occur in different regions? For concerted changes within a single species, the answer is gene flow.

At this point, you might be thinking that there is a basic paradox here. If gene flow is necessary to maintain all populations within the same species, then how can regional characteristics be maintained? It might seem that gene flow would eliminate geographic differences, and the persistence of regional continuity would argue for no gene flow. It would seem at first glance that you can't have it both ways. Too much gene flow and there would be no regional traits. Too little and you can't keep all populations in the same species. Wolpoff and colleagues developed the model of multiregional evolution to explain this seeming paradox. The essence of their argument is that there is a *balance* between evolutionary forces that simultaneously maintain species integrity while also allowing for the maintenance of regional characteristics. They further note that natural selection also affects this balance in the case of adaptive traits.

It is not unusual to see statements debating multiregional evolution because of the supposed paradox, but the concept of balance in evolutionary forces is well known to population geneticists. The arguments put forth by Wolpoff and colleagues are amply documented in their papers and books, and I will not repeat their exact thoughts here. Instead, I have devised a different way for anthropologists to understand the dynamics of balance in the multiregional evolution model. Given my own background, I tend to understand the concepts best in terms of simple mathematical models and computer simulations, the latter of which is used here. The simplest case is a balance between gene flow and genetic drift. Gene flow will act to make populations more similar *on average*; that is, averaged across all loci. Genetic drift, however, will cause individual loci to deviate from this expected average. By chance, some loci will deviate only slightly, whereas others will deviate quite a bit. In terms of the evolution of a population under the simultaneous action of gene flow and genetic drift, this means that *some* loci will continue to show regional continuity even though the average of all loci reflects populations becoming more homogeneous.

Because the balance between gene flow and genetic drift is not always intuitively obvious, a simple simulation helps make these basic points, as shown in Figs. 3.9, 3.10, and 3.11. I looked at the changes in the frequency of an allele in two populations, each consisting of 500 reproductive adults, exchanging one migrant each generation. I set the initial allele frequencies to 0.1 for population 1 and 0.9 for

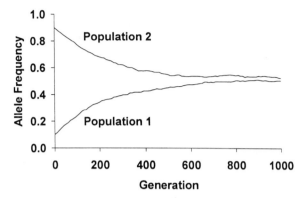

Figure 3.9. *Computer simulation results illustrating the interaction of gene flow and genetic drift. The simulation starts with initial allele frequencies of 0.1 for population 1 and 0.9 for population 2. Each population consists of 500 reproductive adults, and the two populations exchange one migrant each generation. Genetic drift and gene flow are simulated for 1000 generations. This graph shows the average allele frequencies over 200 separate simulations to give an idea of the average evolutionary changes. When averaged over 200 separate simulations, the allele frequencies become increasingly similar over time, showing the impact of gene flow. The individual simulations do not all resemble the average pattern shown here, as illustrated in Fig. 3.10.*

population 2 to simulate a case of high initial populational differences. I then simulated gene flow and genetic drift for 1000 generations. Because genetic drift is a random process, each time the simulation is run, different results could occur. To investigate the average pattern, and the variation around this average, I repeated the simulation 200 times. Figure 3.9 shows the allele frequencies for each generation in both populations *averaged* over all 200 runs. The pattern is very similar to that expected from gene flow (as shown earlier in Fig. 2.8)—the two populations become increasingly more similar to each other over time. The curves aren't completely smooth because of the effect of genetic drift (if more runs were used, the curve would become smoother). After 1000 generations, the average allele frequencies are virtually the same (population 1 = 0.507, population 2 = 0.526).

Because genetic drift is a random process, each of the individual 200 runs is likely to be different. Figure 3.10 shows four separate runs, each showing a different pattern of allele frequency change. Figure 3.10a is not that different from the average expectation; although the allele frequencies in both populations fluctuate, the overall pattern is one in which the two populations become more similar to each other over time. Figure 3.10b shows a different pattern, in which both populations wind up with final values most similar to the initial values in population 2. Figure 3.10c is an example of the opposite pattern, and both populations wind up most similar to the initial values in population 1. Figure 3.10d shows yet another pattern, with both populations fluctuating over time but each winding up after 1000 generations with values not far from where they started. If these were actual allele frequencies and we could compare the initial values with those after 1000 generations, what kinds of patterns would we see in terms of regional continuity? Population 1 shows high continuity in Figs. 3.10c and 3.10d, because the final allele frequencies are similar to where they started. Population 2 shows high continuity in Figs. 3.10b and 3.10d for the same reason.

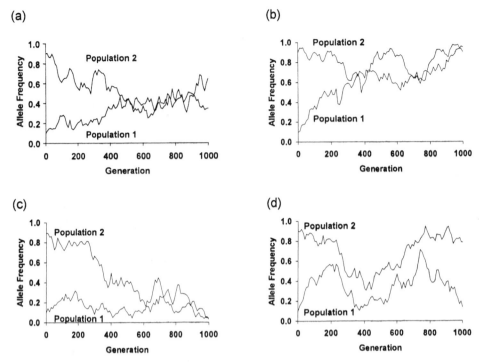

Figure 3.10. *Four examples (out of 200 separate runs) of the simulation described in Fig. 3.9. Note that individual simulations often give results different from the average pattern shown in Fig. 3.9.*

We see the powerful influence of genetic drift on variation of individual loci from the average expected values, something I think is often ignored in analyses of human evolution. However, this variation poses a problem—how much regional continuity should we expect to see under the multiregional model? The way to answer this question for any particular set of parameters is to look at the entire distribution. Figure 3.11 shows the distribution of allele frequencies for population 2 after 1000 generations. This graph plots all of the individual 200 runs used in the simulation. There is considerable variation around the average value; some runs produced final allele frequencies very similar to the initial values (high continuity), whereas others showed quite different values (low continuity). Most final values are closer to one extreme or the other, with fewer individual frequencies close to the average. This U-shaped distribution is familiar to population geneticists; it results when there is considerable genetic drift. The bottom line of this simulation is that we should *not* expect *all* traits to show regional continuity under multiregional evolution, nor should we even expect *most* traits to show high continuity. We should expect, however, to see *some* traits with high continuity; the exact proportion of these traits depends on the specific balance between gene flow and genetic drift, which is affected by population size and migration rate. In her comprehensive analysis of continuity traits in East Asia and Australia, Marta Lahr found evidence of continuity in 11 out of 30 traits.[29] She further argued that because most traits did not show continuity, the multiregional model was incorrect. I argue that we should *not* expect to see overwhelming continuity under multiregional evolution and that her

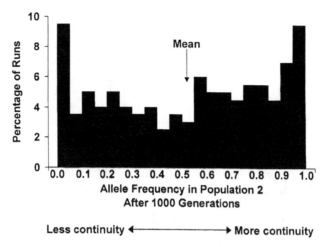

Figure 3.11. *Histogram of the final allele frequencies in population 2 for 200 simulation runs. The mean value corresponds to the final allele frequency in Fig. 3.9. Individual runs vary considerably in terms of their final allele frequencies; some wind up very low (little continuity), whereas others wind up high (showing more continuity). These results, combined with those from Fig. 3.9, show that under a model of gene flow and genetic drift, the two populations become more similar over time on average, whereas some individual loci will maintain high continuity because of genetic drift.*

results are consistent with a multiregional interpretation. On the other hand, we should not expect to see much continuity under a replacement model, a point I return to in Chapter 9.

3.4.2 The Location and Timing of the Transition—Africa or Worldwide?

The debate over the mode of the transition is actually fairly simple—was it a speciation event or evolution within a species? I think that some of the confusion over different modern human origins models lies in confusing the mode of the transition with predictions about the location and timing of the transition. There are two competing hypotheses regarding where and when modern humans first appeared. One is the idea of a recent African origin of modern populations; whatever genetic change(s) occurred in the transition from archaic to modern humans occurred first in Africa between roughly 100,000 and 200,000 years ago. Although this idea is basic to the idea of African replacement, it can also be accommodated within a multiregional perspective. Under replacement, the new modern population is a new species that later disperses out of Africa to replace archaic humans. Under *some* multiregional models, this new population is not a new species but part of a geographically widespread single species. The genetic changes that began in Africa are then shared with the rest of the species through genetic exchange in one of three ways: 1. Gene flow between relatively stationary populations, 2. Long-distance migration with ensuing admixture, or 3. Both. In the first scenario genes flowed from one population to the next through the normal exchange of mates. A second scenario invokes the actual movement of populations out of Africa that subsequently mate with archaic populations. Finally, it is also possible that *both* isolation by distance and dispersal with admixture occurred.

An alternative to a recent African origin of modern humans is the idea that there was no single time or place for the emergence of modern human populations. This view, championed by Wolpoff and many of his colleagues, states that there was no single *populational* origin of moderns, but instead there were multiple origins for modern *anatomic characteristics*. Under this view, *some* modern traits appeared in Africa, but *others* appeared first in other regions. Gene flow between populations led to the spread of these new features throughout the Old World, resulting in a gradual shift toward modernity as they became more common throughout the species. Modern humans, therefore, did not arise in a single event but instead came about through the coalescence of modern traits appearing throughout the species' range. The spread of certain features (e.g., the gracility in early Africans) may have been promoted by natural selection. Neutral traits might also have become more common because of differences in population size, with certain regions (Africa) having greater genetic impact because of their larger size (this point is returned to in Chapter 5). According to Wolpoff and colleagues, this regional coalescence, unlike a singular origin of modern humans, explains why we have so much difficulty in defining the exact boundaries between archaic and modern human fossils—because there is no single origin event the transition is gradual, and any distinction between archaics and moderns during the transition is arbitrary.

The difference between a recent African origin of modern *populations* and a regional coalescence of modern *traits* is shown by comparing Figs. 3.12 and 3.13. Both figures represent a very simplified view of the evolution over time in three regional populations, represented by rectangles. Three different traits are represented by the circles, squares, and triangles within each box. For simplicity, each trait is described as having one of two different states—"archaic," represented by open shapes, and "modern," represented by shapes that have been filled in. Figure 3.12 represents a single population origin for modern traits. It starts with each of the three populations all having the "archaic" condition. In the next time period, there has been a change from the "archaic" to "modern" condition in the middle population, representing a recent African origin. In the last time period, the modern traits have spread to the other regions, either through replacement of one species by another or gene flow between populations within the same species. Figure 3.13 shows the pattern of the regional coalescence of traits. The starting and ending points are the same as in Fig. 3.12, but the process is different. During the second time period, each regional population is marked by the shift from "archaic" to "modern" in only one of the three traits. During the last time period, these changes have coalesced through gene flow, and all regional populations are "modern." The key difference between a recent African origin and regional coalescence is that in the former most (if not all) of the evolutionary change from archaics to moderns takes place within Africa, whereas in the latter these changes take place in different places for different traits and modernity is reached by the coalescence of all these changes within a single species.

3.4.3 Evolutionary Models

My attempt to synthesize arguments about modern human origin models is based on considering both of the aspects discussed above: 1. The mode of the transition

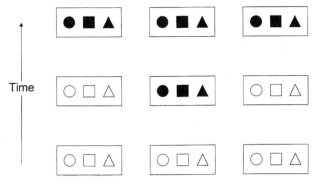

Figure 3.12. *A simplified schematic diagram illustrating a recent African origin of all traits. The rectangles indicate three regional populations evolving over time. The circles, squares, and triangles within each rectangle illustrate three separate traits. For simplicity, each trait is shown in two possible states—"archaic" (unshaded) and "modern" (shaded). The changes from archaic to modern all take place within one region (center time period) and then spread to the other regions (top time period). This model corresponds to the development of all or most modern traits within a single region (Africa), from which these changes then spread throughout the remainder of the Old World. This general model does not specify how the change takes place, either through replacement or through gene flow. Contrast this model with the one shown in Fig. 3.13, where the changes from archaic to modern take place in different regions.*

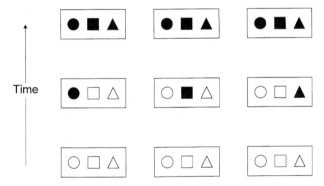

Figure 3.13. *A simplified schematic diagram illustrating multiregional coalescence of changes from archaic to modern traits. The rectangles indicate three regional populations evolving over time. The circles, squares, and triangles within each rectangle illustrate three separate traits. For simplicity, each trait is shown in two possible states—"archaic" (unshaded) and "modern" (shaded). Different changes take place within different regions (center time period). All of these changes coalesce throughout all regions because of gene flow over time (top time period). Contrast this model with the one shown in Fig. 3.12, where the changes from archaic to modern all take place within a single region.*

and 2. The location and timing of the transition. These two aspects are diagrammed simultaneously in Fig. 3.14, which shows three general classes of models that I use throughout the remainder of this book.[30] I define these models as follows:

• *The African Replacement Model.* Modern humans began as a new species in Africa between 100,000 and 200,000 years ago. By 100,000 years ago, some populations dispersed out of Africa, replacing the archaic populations outside of Africa. All of our ancestors 200,000 years ago lived in Africa.[31]

- *The Primary African Origin Model*: The transition from archaic to modern humans began in Africa between 100,000 and 200,000 years ago. These genetic changes were later shared throughout the rest of the Old World, gradually transforming all archaic populations into modern populations. Different variants of this model stress the relative importance of gene flow between populations and dispersal with admixture in spreading the genetic changes across the species.

- *The Regional Coalescence Model*: Different traits underwent a transition to modernity in different regions at different times and were later spread throughout the species by gene flow. Modern humans gradually appear because of the coalescence of these changes throughout the Old World.

The way I have set up Fig. 3.14 shows that there is overlap between these general models. Both the African replacement model and the primary African origin model share a recent African origin for modern traits. In addition, both the primary African origin model and the regional coalescence model share multiregional evolution (within a species) as the mode of evolutionary change. Note that one box in the figure is left blank. By definition, speciation requires a single origin point in time and space, and therefore there is no model that incorporates speciation and multiple regional origins.

Part of the confusion in the literature comes from using the same names for different parts of the diagram in Fig. 3.14 (I'm not assigning blame—I have unfortunately done the same thing far too frequently). For example, the term "recent African origin" refers sometimes to the general location and timing of the transition and other times to the specific model of speciation and replacement. The term "multiregional" is also used in different contexts. In some cases, "multiregional" is used to refer to the general model of multiregional evolution, which refers to evolutionary changes within a geographically widespread species. In other cases, the term has referred more specifically to the regional coalescence model. *Both* the primary African origin model and the regional coalescence model are by definition

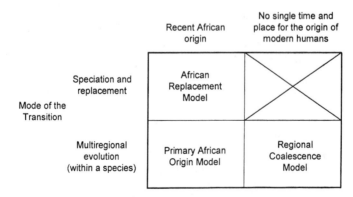

Figure 3.14. *Summary of models of modern human origins. Models are classified according to two features: 1. Mode of the transition (speciation versus multiregional evolution), and 2. Location and timing of the transition (recent African origin versus regional coalescence).*

multiregional evolution models, a point made very clear by Wolpoff and colleagues in a 1994 book chapter:

> Several different patterns of *modern* human origins are compatible with the Multi-regional Evolution model. For instance, it is possible that modern humans originated as a single population that spread and hybridized with other populations, with the most successful of the "modern" traits spreading furthest, perhaps even beyond the range over which the hypothesized first population can be identified. This would certainly fit the modeling of evolutionary change across the polytypic species in which the changes are caused by shifting balances of genic exchange and selection (or drift) magnitudes. One of us (FHS) is inclined toward this explanation. The rest of us think this is most unlikely, and believe that the best reading of the fossil record suggests a different compatible pattern—that modern humans evolved through the coalescence of a series of modern traits that appeared independently in various areas at different times.[32]

This quote shows that we must draw a contrast between the *general* process of multiregional evolution and the *specific* evolutionary hypotheses compatible with this general model.

Some discussions of modern human origins models draw a contrast between the African replacement model and the specific multiregional model of regional coalescence and describe other models as "intermediate." This is not entirely correct, because these other models are multiregional in the most general sense. One example is the "assimilation model" proposed by Fred Smith,[33] who argues that the initial evolutionary changes leading to modern humans took place in Africa and then spread throughout the remainder of the species through the process of gene flow between neighboring populations. Another example is the "Afro-European hybridization model"[34] proposed by Gunter Bräuer, who also suggests that the initial evolutionary changes leading to modern humans took place in Africa and then spread to populations outside of Africa. In this case, however, the mechanism of genetic spread occurred through the actual movement of populations dispersing out of Africa and mating with non-African populations. I see the key difference between these models as the mechanism of genetic exchange (gene flow through isolation by distance or admixture) and the magnitude of the genetic exchange. However, in a series of articles in the journal *American Anthropologist*, Bräuer identifies himself as an advocate of "recent African origin" and Smith aligns himself with "multiregional evolution."[35] I would argue that both of these models are multiregional in the general sense of involving genetic input from outside of Africa and that both are specifically primary African origin models.

Of course, there are variants of the three models outlined in Fig. 3.14. The example outlined above comparing the models of Smith and Bräuer is one example. It is also possible to describe models that incorporate pieces of the speciation and multiregional models. Suppose, for example, that all human evolution over the past two million years was multiregional except for the Neandertals, which represent a different species that were later replaced by modern populations from elsewhere. In this case, we can imagine multiregional evolution operating in some parts of the Old World but not everywhere. We could further argue that multiregional evolution in the non-European part of the Old World resulted from either a primary African origin or regional coalescence. There are obviously many possible models and submodels, but further elaboration requires that we resolve the basic issue of speciation versus multiregional evolution and the location and timing of the transition. The next chapter turns (finally!) to considering what the genetic evidence can (and cannot) tell us about these questions.

3.5 SUMMARY

The general picture of evolution within the genus *Homo* is clear. Early humans (*Homo erectus*) evolved in Africa roughly two million years ago and then began dispersing across the Old World. Over time, these early humans evolved into larger-brained "archaic" humans and then later into "anatomically modern" humans. We also have a good general picture of cultural changes occurring over the past two million years, including development and refinement of stone tool technology, the adoption of a hunting and gathering lifestyle, use of fire, and the development of symbolic expression, such as cave burials and prehistoric art.

There is disagreement about the fine details and underlying mechanisms of these general changes. Perhaps the most controversial discussion centers on the question of the number of species, ranging from views that see only one species at any point in time to those incorporating five or more separate species. The debate over the number of species is a major concern in discussions over the origin of anatomically modern humans. Was the mode of transition speciation (and subsequent replacement outside of Africa) or multiregional? The debate also concerns the location and timing of the transition to modern humans—was it a single region origin or did it result from the regional coalescence of modern traits?

For several years, I have read and reread these debates and have often been confused over the specific statements and predictions of various origin models. Furthermore, I think that this confusion has been even more problematic in the genetics literature (and, to be fair, I should note that I have probably contributed to the confusion as well!). My attempt here is to describe the debate over modern human origins in terms of three models: African replacement, primary African origin, and regional coalescence. The latter two, however, are variants of the general multiregional model. Most of the discussion throughout the next five chapters will focus on the contrast between the speciation/replacement model and the general multiregional model (dealing with genetic input from at least two geographic regions that took place within a single species).

Chapter **4**

In Search of Our Common Ancestor

One of my interests is in family genealogy. I am fortunate that relatives in both of my parents' families had similar interests and passed along to me much valuable information to build upon. Anyone interested in genealogies knows the many difficulties in reconstructing the past, including lack of records, name changes, migrations, and other problems. Even when we find a remote ancestor, we seldom know much about the person other than his or her name and ages of birth, marriage, and death. In many cases, we don't even have that much information, as in the case of several remote relatives for whom I know only the first name of their spouse. Still, it is fascinating to recover something of your own past, giving history more of a personal feel.

Reconstructing a genealogy involves identifying an increasing number of ancestors in each previous generation. You have two ancestors one generation back—your parents. You have four ancestors two generations back—your grandparents. Three generations back you have 8 ancestors, four generations back you have 16 ancestors, and so forth. The number of ancestors follows a simple mathematical relationship—the number of ancestors is equal to 2^t, where t is the number of generations. This is actually a maximum number of unique ancestors, because some of your ancestors may be the same person. For example, imagine that your father's father's father's father's father's father is the same person as your mother's father's father's father's father's father. This means that you would actually have only 63 ancestors six generations ago rather than 64, and your parents would be half-fifth cousins. At some point in time, we are all interrelated distant cousins, often in several ways. The mathematical logic of this statement is straightforward. For example, we can compute the maximum number of ancestors you had only 40 generations ago (roughly 1000 years) as $2^{40} = 1,099,511,627,776$, which is more people than lived at the time (or even today—this figure is over 180 times the current human population of the world). In most cases, we can only identify a small handful of our ancestors (in my case, my oldest information is on two relatives 11 generations ago, whereas there could be a maximum of 2048 unique ancestors that far back). Although interesting, constructing a family tree from historical records is difficult at best. The farther back in time we go, the worse things get because of the increasing number of ancestors and lack of records.

Genetic data can also be used to reconstruct genealogies of genes, which can be used to make inferences about population history. The method begins with the comparison of DNA sequences from different individuals. Sampling is done from different species if the goal is to produce a gene tree that is used to infer phylogeny or from different individuals within a species if the goal is to produce a gene genealogy for that species. We start with the principle that genetic similarity reflects common ancestry and look for patterns of similarity that reflect past events. This is not as simple as it sounds, because genetic similarity can reflect different factors, not all of which can relate to history. We look for traits shown to be (or strongly assumed to be) neutral so that we don't have to deal with complications due to natural selection. Analysis of nuclear DNA is further complicated by the problem of multiple ancestors. A given DNA sequence could come from your mother or your father, who in turn could have inherited it from their mother or father. Which grandparent contributed the DNA sequence? This question becomes harder and harder to answer as we consider even more remote ancestors. For this reason, we are primarily interested in DNA that is inherited without recombination. The first, and perhaps the most famous, approach used mitochondrial DNA.

4.1 "MITOCHONDRIAL EVE"

One of the most famous applications of genetic methods to genealogical reconstruction was the paper by Rebecca Cann, Mark Stoneking, and Allan Wilson. Their paper, entitled "Mitochondrial DNA and human evolution," appeared in 1987 in the prestigious British journal *Nature*. Although some earlier papers on paleoanthropology had proposed the idea of a recent African origin and replacement, the publication by Cann et al. was a major milestone in the rapidly developing debate over modern human origins with the development of the idea that *all* modern humans share a common female ancestor who lived in Africa roughly 200,000 years ago. The idea of a single female ancestor is similar enough to the Biblical story of Adam and Eve for it to be almost irresistible that this common ancestor would be given the name "Eve."

4.1.1 Analyzing Mitochondrial DNA

As noted in Chapter 2, mitochondrial DNA (mtDNA) is invaluable in evolutionary analyses because it is inherited maternally. The entire mtDNA sequence (haplotype) is inherited without recombination through the mother's line; in any given generation, you have only *one* ancestor. Your mtDNA came from your mother, who got it from her mother, and so on back into the past. At any generation in the past, you have only one mitochondrial DNA ancestor, whereas you have thousands of nuclear DNA ancestors even a dozen generations ago. Looking at mtDNA provides an easier way to make inferences about common ancestors. Figure 4.1 shows the genealogy of two people, Andy and Alice, for three ancestral generations. They each

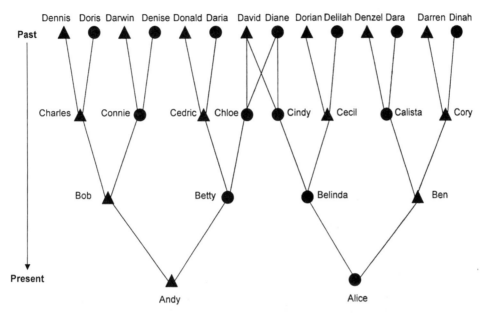

Figure 4.1. *Hypothetical genealogy for two related individuals (Andy and Alice) who are second cousins. They share one set of great-grandparents (David and Diane), who are their common ancestors three generations in the past. Triangles indicate males, and circles indicate females.*

have two parents, four grandparents, and eight great-grandparents. Note that they also share a set of great-grandparents, making Andy and Alice second cousins. They therefore have two common ancestors, great-grandfather David and great-grandmother Diane. Any genetic similarity due to inheritance from a common ancestor could have come from David or Diane.

Contrast this genealogy with the mitochondrial DNA genealogy of the same individuals in Fig. 4.2. Even though Andy and Alice have eight ancestors three generations back in the past for their nuclear DNA, they have only one mitochondrial DNA ancestor in each generation. In addition, Andy and Alice share only one common mitochondrial DNA ancestor—great-grandmother Diane. Looking at this another way, we can say that the most recent common mitochondrial DNA ancestor of Andy and Alice is Diane, their mother's mother's mother.

Now, consider this genealogy as part of a larger genealogy that has two other second cousins, Albert and Anna, who have great-grandmother Danielle as a common mitochondrial DNA ancestor. Further, imagine both that Diane from Fig. 4.2 is the first cousin of Danielle and that they both have a common mitochondrial DNA ancestor in their grandmother Frances. This mitochondrial genealogy is shown in Fig. 4.3, where Frances is the most recent common mitochondrial DNA ancestor of Andy, Alice, Albert, and Anna. We could imagine additions to this scenario, linking increasingly distant cousins to an increasingly distant common ancestor.

At some point in the past, any two individuals will have a common mtDNA ancestor. If we add a third person to the analysis, all three will share a common mtDNA ancestor at some point in the past. If we extend this principle to all living humans,

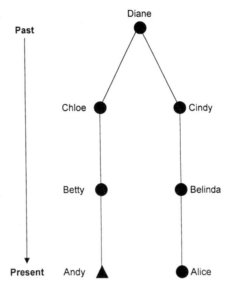

Figure 4.2. *The mitochondrial DNA (mtDNA) genealogy for Andy and Alice (see Fig. 4.1). Because mtDNA is inherited only through the mother's line, Andy and Alice have only one ancestor in any given generation and only one common ancestor three generations earlier—Diane. Note that in subsequent generations Andy's mtDNA will be lost because he is male and cannot pass his mtDNA to his children—they will inherit the mtDNA from Andy's mate(s).*

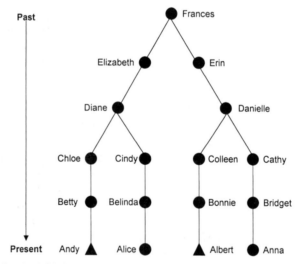

Figure 4.3. *Mitochondrial DNA genealogy linking two sets of second cousins (Andy/Alice and Albert/Anna) through their most recent common ancestor, Frances.*

it means that we all share a common mtDNA ancestor at some point in the past. To understand why this is the case, we must consider the likelihood of extinction of certain mtDNA lines. Assume that a female with a particular mtDNA sequence existed at some point in the past. This female might not have reproduced, in which

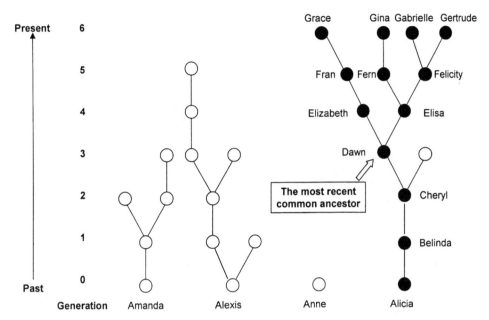

Figure 4.4. *Simulated extinction of mitochondrial DNA. The simulation begins with four females (Amanda, Alexis, Anne, and Alicia), each of whom has two children. A random number generator was used to pick one of three possible distributions for each pair of children: Two boys (probability = 0.25), one boy and one girl (probability = 0.50), or two girls (probability = 0.25). Because mitochondrial DNA is inherited through the mother's line, only females are shown in this figure. By chance, Amanda had one girl, Alexis had two girls, Anne had no girls, and Alicia had one girl. Anne's mitochondrial DNA was then extinct. The simulation was then repeated again for six generations, by which time there were four females that carried some of the mitochondrial DNA from the starting generation—Grace, Gina, Gabrielle, and Gertrude. Tracing back, we see that these four females share several common ancestors—Dawn, Cheryl (Dawn's mother), Belinda (Dawn's grandmother), and Alicia (Dawn's great-grandmother). Going back in time from the present (generation 6), we see that Dawn is the most recent common ancestor (MRCA).*

case her mtDNA would become extinct. Alternatively, she might have had only sons, in which case the line would also become extinct, because mtDNA is inherited only through the female line. On the other hand, she might have had all daughters, in which case her mtDNA would be amply represented in the next generation.

Figure 4.4 illustrates this principle by starting with four hypothetical female ancestors—Amanda, Alexis, Anne, and Alicia—and tracing their genetic contributions over several generations. We then conduct a simple experiment in probability. Let us assume that each female has two children. Each child could be male or female, depending on which sex chromosome is inherited from the father. There is a 50% chance of having a boy or a girl. The probability of having *two* boys is $1/2 \times 1/2 = 1/4$. The probability of having two girls is also $1/4$. The probability of having one boy *and* one girl (in either order) is $1/2$. By picking random numbers, we can simulate the chance of these events and trace the effect on the inheritance of mtDNA.

Let us examine the results of the simple simulation shown in Fig. 4.4. In the first generation, both Amanda and Alicia had one daughter each, Alexis had two daughters, and Anne had no daughters. From that point on, Anne's mtDNA has been lost even though she would continue to contribute genetically to future generations through the inheritance of her nuclear DNA by her sons. By the third generation, Amanda's mtDNA has been lost by chance. By the fifth generation, Alexis' mtDNA has also been lost by chance. All of the people represented in the sixth generation (Grace, Gina, Gabrielle, Gertrude) trace their mtDNA back to a common ancestor in the third generation (Dawn), who in turn inherited her mtDNA from Alicia. All four individuals in the "present" trace their mtDNA back to a single female— Alicia—who lived six generations earlier. In this case, Dawn represents the most recent common ancestor (MRCA). This example is somewhat unrealistic because I only considered the relative probability of 0, 1, or 2 daughters. In reality, we would have to factor in differences in reproduction. However, the basic principle is the same—that all mtDNA in living people traces back to a single female ancestor. Because we don't actually know the true genealogy for more than a few generations (depending on the extent and quality of written records, if any), we can use genetic data to make inferences about the most recent common ancestor (Dawn in Fig. 4.4).

The use of mtDNA certainly simplifies things quite a bit. However, it also introduces a seeming paradox. If a set of mtDNA sequences ultimately traces back to a single ancestral female, then we might expect that everyone has the same exact same mtDNA. In other words, we would all be identical. If so, then how can mtDNA be used to untangle relationships among individuals? If everyone has the same mtDNA, then there is no way to tell which individuals are more closely related. We would not be able to tell, for example, that Gertrude and Gabrielle were more closely related to each other than either is to Grace.

The answer is mutation. We expect that as mtDNA is passed along from generation to generation through the maternal line occasionally a mutation will occur, which will then be passed on to the next generation. Over time, more mutations will accumulate. This principle can be used to make inferences about common ancestry. Unlike the examples in Figs. 4.1–4.4, we don't know the underlying genealogy; instead, we attempt to reconstruct it by examining genetic differences between pairs of individuals and finding out how many mutational differences have accumulated. In general, the greater the length of time separating two individuals, the more mutations will accumulate, and the greater the genetic difference between them. This is the same basic idea described in Chapter 2 for looking at the genetic difference between human and chimpanzee species. The approach is phylogenetic, in that it is concerned with the reconstruction of ancestry using a model of accumulating mutations over time. Other methods derived from a field known as coalescence theory are described later in this chapter.

As an example of the phylogenetic approach, Fig. 4.5 considers a very simple case in which we look at mtDNA sequences from three different people: Alice, Betty, and Cindy. We first look at all possible trees that have two of these people more closely related to each other than to the third. There are three possible trees:

Alice and Betty are more closely related to each other than to Cindy.
Alice and Cindy are more closely related to each other than to Betty.
Betty and Cindy are more closely related to each other than to Alice.

Imagine three people. Alice has nucleotide A, Betty has nucleotide G, and Cindy has nucleotide A. If we know that the common ancestor of all three women had nucleotide G, there are three possible trees:

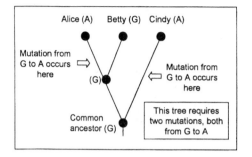

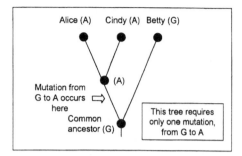

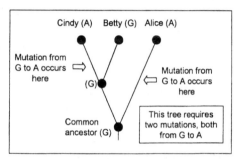

Figure 4.5. *An example of reconstructing a genealogy from genetic data. For a given nucleotide, the three women have a different base: Alice and Cindy have base A, whereas Betty has base G. There are three different ways of linking these three women in a genealogy. Trees (a) and (c) each require two mutation events, whereas tree (b) requires only one mutation event. Tree (b) is therefore considered the most parsimonious.*

Which of these three genealogies is correct? (For technical reasons, we rarely have to worry about the other possibility, that all three are equally related to each other). Imagine we look at one nucleotide site and find that both Alice and Cindy have base A, whereas Betty has base G. Because there are two bases, we might assume that at some point in the past there was at least one mutation, perhaps from A to G, or perhaps from G to A. Another possibility might be that both mutated from yet a third base, but we in general try to focus our efforts on the most parsimonious explanation that requires the fewest mutations.

To build a "tree" we need to know who the common ancestor of all three people was, so that we can determine what direction a previous mutation might have taken. There are several ways of doing this; the most appropriate method is to look at the DNA sequence of an "outgroup"—another sample related more distantly to the subjects. When looking at large numbers of humans a typical outgroup is the chimpanzee. In this hypothetical case, let us assume that the outgroup comparison shows us that the ancestral base was G.

We can now look at each of the three possible trees in Fig. 4.5. The first tree suggests that Alice and Betty had a common ancestor, who in turn had a common ancestor with Cindy. Because we know that the ancestral form was G, this means that this tree would require at least two separate mutations (both of which would be from

G to A). The second tree requires only one mutation, and the third tree requires at least two mutations. We look for the most parsimonious explanation—which is more likely, one mutation or two identical mutations? Parsimony dictates the simplest explanation, which is the one mutation as required by the second tree in Fig. 4.5. In this tree, the common ancestor had G. Betty inherited this base from the common ancestor. A mutation then occurred from G to A in the common ancestor of Alice and Cindy, who both inherited A from this common ancestor.

Of course, we would not have much reliance on a genealogy based on a single nucleotide site. We need a much larger sampling of DNA sequences to come up with the most likely genealogy. As we sample more and more of the genome, we are likely to run into specific cases that don't fit. For example, try repeating the example in Fig. 4.5 with a different section of DNA where Alice has base G, Betty has base G, Cindy has base T, and the outgroup comparison shows that the common ancestor had base T. The most likely tree, linking Alice and Betty to a common ancestor, is the most parsimonious. However, this parsimonious tree is different from tree (b) in Fig. 4.5. Which tree do you pick, the one based on the first set of data (Fig. 4.5) or the second?

Computer methods are needed for tree reconstruction in almost all cases. The process gets even more complicated if we look at many sites for large samples of DNA sequences because we will not be able to look at all possible trees. In the above example, we looked at three different DNA sequences, one from each person. For three sequences, there are three possible trees. For four sequences, the number of possible trees is 15. The number of possible bifurcating trees (with only one split at any point in time) increases very quickly, as shown here:

Number of DNA sequences	Number of possible trees
3	3
4	15
5	105
6	945
7	10,395
8	135,135
9	2,027,025
10	4,459,425
20	8.2×10^{21}
50	2.8×10^{76}

Because some studies use over a hundred DNA sequences, the total number of possible trees is astronomical. For example, Linda Vigilant and colleagues looked at 135 mtDNA sequences, which results in excess of 10^{267} possible trees.[1] Within any reasonable sample size, we will not be able to look at all possible trees because the number is too great for even the fastest computers currently available. Instead, sophisticated computer programs have been developed to statistically search subsets of possible trees. The most commonly used program is PAUP (Phylogenetic Analysis Using Parsimony) written by David Swofford of the Illinois Natural History Survey.

There are further complications, such as multiple substitutions (also known as multiple hits) that occur when a single site has more than one mutation. For example, consider a nucleotide sequence of CAGT in a common ancestor. Imagine that a mutation from A to C occurs in the second base in one descendant line and a mutation from A to T occurs in the second base of another descendant line. If we look at the two descendants we would find the sequence CCGT in one and CTGT in the other. If we just compared these two we would see only one difference—a C versus a T in the second position. We would then infer that only one mutation had occurred (either C to T or T to C) when in fact two mutations had occurred, both at the same site. As a result, we would underestimate the true genetic difference between the two lines. Standard computer programs (such as PAUP) have ways of handling this type of problem.

Once a gene tree has been reconstructed, a next useful step is to estimate *when* a common ancestor lived. A gene tree gives us an estimate of how much genetic change has taken place over time. Given this information and an estimate of the average rate of mutation, we can estimate the number of generations back to the common ancestor. This is essentially the same method of molecular dating as described in Chapter 2 except that it is applied here to genetic lines within a species. As a hypothetical example, consider a gene tree that shows an average of 0.45% sequence divergence between living people and the root of the gene tree. To translate this divergence into actual time we need to have an idea of the rate of change due to mutation. In this case, let us assume that the rate is 3% per 50,000 generations, which equals 0.00006% per generation. To estimate the date for the common ancestor, we divide the sequence divergence by the rate, giving $0.45/0.00006 = 7,500$ generations. If we then assume roughly 20 years per generation, this figure translates into $7,500 \times 20 = 150,000$ years ago. As with any molecular estimate we need to have the best possible estimate (or range of estimates) available for calibration. In addition, we have to be aware that such estimates are likely to have very wide margins of error, as is described later in this chapter.

4.1.2 The Analyses

Given this background, we are now in a position to examine more closely the original paper by Cann, Stoneking, and Wilson.[2] Cann and colleagues collected mitochondrial DNA sequence data from 147 people. Most of these sequences (145) were derived from placentas obtained from U.S., Australian, and New Guinean hospitals. At the time, placental tissue was an important source of mtDNA. Since then, new genetic methods have allowed extraction of mtDNA from other sources, including plucked hairs.

Each person's maternal ancestry was evaluated from an interview and used to place every one into one of five major geographic regions depending on ancestry: Sub-Saharan Africa, "Caucasians" (representing Europe, North Africa, and the Middle East), East Asians, aboriginal Australians, and aboriginal New Guineans. Only indigenous people were sampled for the latter two regions to get a more accurate picture of human variation before the widespread migrations of the past 500 years. This restriction avoids the problems that would arise if Australians whose recent ancestors had moved from Europe were included; to get the best estimate for ancient Australians, the aboriginal population had to be sampled.

Restriction mapping (as described in Chapter 2) was used with 12 restriction enzymes, providing an average of 370 restriction sites per person. Some of the 147 mtDNA sequences were identical and therefore removed from the analysis, leaving 133 unique mtDNA types. The PAUP computer program was then used to construct a tree of mtDNA lines. Because there were no comparable data for chimpanzees Cann et al. could not use the outgroup method to locate the root of the tree. Instead, they used a method known as midpoint rooting, where the root is estimated as the average between the two most divergent lines. The final tree (Fig. 4.6) shows two main clusters; one consists only of individuals with sub-Saharan ancestry, and the other contains individuals from all geographic regions, including Africa.

Cann and colleagues suggested that the most parsimonious explanation for this pattern is that the common ancestor was African. Any other explanation would be more complex, involving many migrations back and forth between continents. The idea of common African ancestry is compatible with a replacement model, but it is also compatible with a multiregional model. All anthropologists agree that the genus *Homo* began in Africa and then later spread out to other continents perhaps as much as 1.8 million years ago. The replacement and multiregional models agree on this, but disagree about what happened next. The replacement model proposes that modern humans arose as a new species in Africa very recently and then spread out to different continents replacing extant hominids. In this view, the origin of modern humans represents another dispersal out of Africa. Both models claim an ancient African origin for *Homo*, but the replacement model claims a second, more recent, African origin. Therefore, the finding of an African root for the mitochondrial tree can fit either model.

Many scientists have suggested that the critical difference between the two models is the date of the most recent common ancestor. The multiregional model suggests a common ancestor far back in the past, whereas the replacement model suggests a much more recent common ancestor. Cann et al. proceeded to estimate the date for the common ancestor by calibrating their tree to estimate the rate of mutation in human mitochondrial DNA. Their calibration method was based on observed sequence differences in New Guinean and Australian populations relative to dates on initial colonization obtained from archaeological data. This method gave an average of 2–4% divergence per million years, similar to rates estimated from a wide range of vertebrate species.

Cann et al. then measured the average amount of sequence divergence from the reconstructed common ancestor (the point labeled "a" in Fig. 4.6) to modern humans; this resulted in an average of 0.57% divergence in mtDNA sequences. Dividing this number by a 2% mutation rate gives a date of 0.57/2 = 0.285 million years, which is 285,000 years. Using the 4% rate gives a date of 0.57/4 = 0.143 million years = 143,000 years. The average of these two estimates is 214,000 years. This date, combined with an African root for the tree, suggested that modern human mtDNA had originated in Africa roughly 200,000 years ago. If this is true, then what happened to human populations that lived outside of Africa before this time? Cann et al. suggested ". . . *Homo erectus* in Asia was replaced without much mixing with the invading *Homo sapiens* from Africa."[3]

It was probably inevitable that someone would eventually give the name "Eve" to human mitochondrial ancestor "a" in Fig. 4.6. In fact, the biblical allusion appears

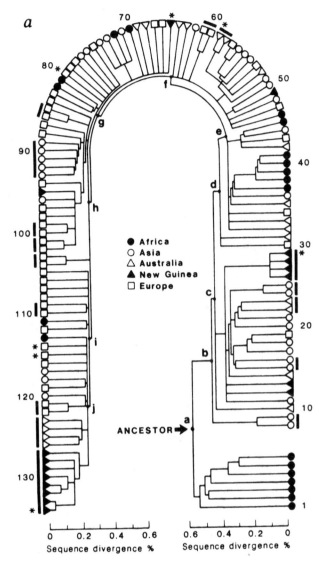

Figure 4.6. *Genealogical tree based on 134 mitochondrial DNA sequences from Cann et al. (1987) (133 sequences as obtained by Cann et al. plus the human reference sequence). The tree shows two major clusters: one consists entirely of sub-Saharan Africans, whereas the other consists of humans from across the world including Africa. The common ancestor, estimated using the midpoint rooting is labeled "a" and is inferred to be African. (from Cann RL, Stoneking M and Wilson A, "Mitochondrial DNA and Human Evolution," Nature vol. 325, p. 464, fig. 3).*

in the very same issue of *Nature* in which the Cann et al. paper appeared, in the form of a news article written by Jim Wainscoat entitled "Human evolution: Out of the Garden of Eden."[4] The name "Eve" soon permeated both scientific and popular media, including the cover story for the January 11, 1988 issue of *Newsweek*, entitled "The Search for Adam and Eve: Scientists Explore a Controversial Theory About Man's Origins,"[5] and a popular book in 1990 by Michael Brown entitled "The Search for Eve."[6]

The focus on a single female common ancestor, although justified in terms of the inheritance of mitochondrial DNA, continues to be misinterpreted, perhaps in part because of the seeming similarity with the biblical idea of a single original pair of human beings. In reality, there were likely to have been at least several thousand (or more) ancestors alive at this time, many of which would contribute genes through time until the present. The difference in perception on the number of ancestors can be illustrated by a simple example using another trait that often is transmitted through the male line—your last name. In some societies, surnames have traditionally been passed down through the male line from father to son. Daughters receive their father's surname but traditionally have taken their husband's surname on marriage (this practice is still common, although declining in recent years). I have 16 great-great-grandparents, but only one had the surname "Relethford." This does not mean that I have only one ancestor four generations ago but that only one contributed my last name. Similarly, only one of these 16 ancestors is responsible for my mitochondrial DNA—my mother's mother's mother's mother. The findings of "mitochondrial Eve" do not mean that we had only one female ancestor 200,000 years ago but instead that only one contributed her mtDNA.

Some anthropologists considered the new evidence on Eve to be confirmation of their ideas on an African origin of modern humans, whereas others questioned the entire approach. Chris Stringer and Peter Andrews published a review paper in the journal *Science* in 1988 that considered the mitochondrial DNA data along with the fossil record and concluded that both are can be explained in terms of a recent African origin.[7] Others, such as Milford Wolpoff, argued that the fossil record did not support a recent African origin and that the mtDNA analysis was therefore flawed.[8]

A number of paleoanthropologists and geneticists questioned the data, methods, and assumptions in the Cann et al. paper. Their study had used African-American subjects to represent sub-Saharan African ancestry, which could be biased because of admixture with people of European ancestry. Because mitochondrial DNA is inherited maternally, this would only be a problem if there were past mating between European-American women and African-American men. Much of the suggested admixture is assumed to have occurred before emancipation and is suggested to have taken place between European-American men and enslaved African-American women. If so, then there would be no bias, because the maternal line would represent African ancestry. However, the existence of admixture suggested a potential problem in terms of biasing the analyses (and therefore the estimated dates).

There were also arguments having to do with technical issues, such as the appropriate choice of mutation rate. Because the date of the common ancestor is determined by dividing the observed sequence divergence by the mutation rate, it is therefore a critical parameter. If a slower mutation rate is used the date will be older, and if a faster rate is used the date will be younger. Others argued that the method used by Cann et al. to estimate the root of the tree (taking the midpoint of the two deepest branches in Fig. 4.6) was inappropriate.

Many of these criticisms were dealt with in a 1991 paper in the journal *Science* by Linda Vigilant, Mark Stoneking, Henry Harpending, Kristen Hawkes, and Allan Wilson.[9] They collected nucleotide sequences (not restriction maps) for 1122 base pairs of the control region of mtDNA, a section of noncoding DNA that is highly variable and shown to mutate rather quickly. These sequences were collected for a sample of 189 people from across the world, including 121 native Africans, 8 African Americans, 20 New Guineans, 1 native Australian, 24 Asians, and 15 Europeans. Most significant was the fact that the African sequences were collected directly on native sub-Saharan Africans, thus avoiding any potential problems of admixture when using African-American samples.

Some of the people in the sample had identical mtDNA sequences, which were excluded, leaving 135 unique types. The PAUP program was used to construct the most parsimonious gene tree. They used chimpanzee mtDNA as the outgroup, a much more efficient method than the midpoint averaging used by Cann et al. Their tree was similar to the earlier one constructed by Cann et al. in showing a primary division between a group of mtDNA types that were all from Africans and a group that contained types from both Africans and non-Africans. The 14 deepest branches (those corresponding to older branches) were all African. The overall tree was confirmation of the Cann et al. tree, providing additional support for an African replacement model and without the potential complications imposed by choice of subjects to represent African ancestry and the method of estimating the root of the tree.

The use of a chimpanzee as the outgroup also allowed a more precise estimate of the age of the human common mtDNA ancestor. After adjusting for multiple substitutions, Vigilant et al. estimated 69.2% sequence divergence between humans and chimpanzees. Given a probable range of four to six million years for the date of the human-chimpanzee split, this figure translates to a divergence rate of $69.2/6 = 11.5\%$ per million years to $69.2/4 = 17.3\%$ per million years. They estimated 2.87% sequence divergence from living humans to the common ancestor. Dividing this number by the estimated divergence rate per million years gave an estimate of $2.87/17.3 = 0.166$ million years using the four million-year calibration date for the human-chimp split and an estimate of $2.87/11.5 = 0.249$ million years using the six million-year date. The estimated age of the common human mtDNA ancestor was between 166,000 and 249,000 years ago. The average of these dates (roughly 208,000 years) is remarkably similar to the figure obtained by Cann et al.

Vigilant et al. viewed their study as further confirmation of a recent African origin. They noted that the recent divergence at about 200,000 years was much too recent to be explained by a multiregional model, which they predicted would require a common ancestor having lived in excess of a million years ago. Their study successfully answered many of the methodological criticisms of the earlier study by Cann et al. Many took this new study as definite proof that modern humans arose relatively recently in Africa, began dispersing out of Africa by 100,000 years ago, and replaced all non-African human populations.

4.1.3 The Debates—Where Did Eve Live?

Debates have continued to this day about the evolutionary significance of "mitochondrial Eve." Some have argued that the above-mentioned studies clearly prove a recent African origin of modern humans with complete (or near complete) replacement of non-African archaic humans. Others argue that the mitochondrial evidence supports a multiregional interpretation or is not conclusive. These debates have focused on the combined findings of an African root and a recent common mitochondrial ancestor. In other words, the debate is over *where* Eve lived and *when* she lived. Different approaches have been developed and applied to the data, often with divergent interpretations.

Soon after the publication of the Vigilant et al. paper in *Science*, doubts were raised about the finding of an African root. The major problem was the method used to estimate the best fitting tree. Because it is not possible to examine all possible trees, the normal method is to examine a random subset. However, Vigilant et al. did not use the correct randomization option, which biased their results. Several studies quickly pointed this out,[10] including one with Mark Stoneking, one of the authors of the 1991 *Science* paper. Reanalysis of the sequence data produced a number of trees that provided a better fit to the data than the one published by Vigilant et al. Some of the new results continued to support an African root, but others did not. The authors of one analysis commented, ". . . the available sequence data are insufficient to statistically resolve the geographic origin of human mito-chondrial DNA."[11] However, a later study by David Penny and colleagues devel-oped a new method for tree estimation that was more efficient at finding true underlying patterns. They applied their method to the Vigilant et al. data and con-cluded that an African root (and origin) was the most likely explanation for human mitochondrial DNA variation.[12]

A recent comparative analysis of mitochondrial DNA casts doubt upon an *exclu-sive* African origin of human mtDNA diversity. Pascal Gagneux and colleagues examined 1158 unique human mtDNA sequences from across the world in com-parison with sequences from African and Asian great apes.[13] Their analysis showed that although the majority of "deep" branches (indicating closer proximity to a common ancestor) were from Africa, there were also several that were from outside of Africa. They conclude that "although the vast majority of the non-African branches of the human clade collapse into a single shallower branch, some deep non-African branches persist. It is not possible to rule out a geographically diverse origin for our species."[14] Again, I would qualify this statement further by repeating that the origin of mitochondrial DNA does not necessarily indicate the origin of our species.

Alan Templeton took another approach to geographic origins in a detailed critique of the Eve hypothesis in a 1993 paper in the journal *American Anthro-pologist*.[15] One of Templeton's main points (one made by many other researchers as well) is that a gene tree does not necessarily imply a population tree. The fact that the mtDNA clusters into two primary groups, African and com-bined African and non-African, does not necessarily mean that the history of the human species has to be described in terms of an African population splitting into African and non-African branches. Templeton noted that geographic associa-tions can arise from several different causes that are not limited to the replacement interpretation of an expanding population out of Africa. Geographic associa-tions are also possible under a model of gene flow and isolation by distance. As such, we need to examine the possibility that the mtDNA gene tree could have been produced by evolutionary factors operating within a geographically wide-spread species.

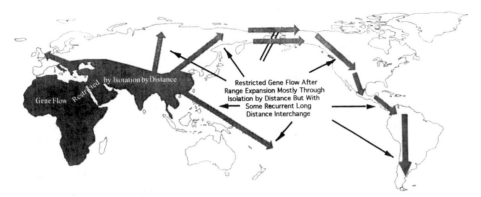

Figure 4.7. *Summary of results from Templeton's geographic analysis of mitochondrial DNA haplotype variation (Templeton 1998, p. 643). This map reconstructs a possible pattern of ancient migration dating back to the coalescence of human mitochondrial DNA roughly 200,000 years ago. Much of Africa and the southern part of Eurasia were characterized by short-range gene flow restricted by geographic distance (isolation by distance). The map also suggests restricted gene flow that followed range expansion in other parts of the world, including Western Europe, Northern Asia, and the New World. (from Templeton AR, Human Races: A Genetic and Evolutionary Perspective, American Anthropologist 100(3), p. 643, fig. 6. Copyright © 1998, American Anthropological Association. All rights reserved).*

To test this idea, Templeton used a method he helped develop that looks at the relationship between mtDNA sequences expected under a geographically based model of gene flow. This method, although complex, is very useful because it allows discriminating between the pattern expected under a model of recurrent gene flow between geographic regions (multiregional) and the pattern expected under a replacement model. The fundamental principle of his method is to combine groups of related mtDNA sequences at different levels of similarity and examine their distribution over time and space.

Templeton's analysis showed that the mtDNA data show evidence of several population expansions but none out of Africa that would cover the entire Old World. Rather, he found that the data are best explained by continued gene flow between geographic regions throughout the past several hundred thousand years. A summary of his findings is shown in Fig. 4.7. Over the past several hundred thousand years, human populations have been connected by restricted gene flow (isolation by distance) throughout Africa, southwest Asia, south Asia, and Southeast Asia. There is also evidence for later expansion into Europe, the Pacific Islands, and the New World. Some of these later expansions could have involved either admixture or replacement. However, even if all of these expansions were replacements, Templeton's analysis suggests that Africa and much of Asia were linked by gene flow throughout much of our species' history, thus rejecting the hypothesis of *complete* replacement. Thus Templeton's analysis supports a multiregional model, although not necessarily one that involves *all* geographic regions. The bottom line of Templeton's work is that a model of gene flow rather than expansion and replacement best explains the mitochondrial DNA gene trees. An African root does not necessarily imply an African origin; Eve had to live somewhere, and it could have been Africa.

4.1.4 The Debates—When Did Eve Live?

As noted earlier, the relatively young date for Eve was a major factor in many people's acceptance of a replacement model. After the work of Cann et al. and Vigilant et al., several other studies of mtDNA also provided support for a recent African origin. Mark Stoneking and colleagues used different methods of tree estimation and derived two estimates of the date of a common ancestor—133,000 years ago and 137,000 years ago.[16] Satoshi Horai and colleagues constructed a gene tree using samples from humans and African apes and estimated a date of 143,000 years for the common human ancestor.[17]

By the mid-1990s, all of these studies were converging on an estimate of a recent African mtDNA ancestor that lived roughly 130,000 to 200,000 years ago, suggesting confirmation of the replacement model. Not everyone agreed with this date. Christopher Wills argued that these studies had failed to adjust properly for variation in mutation rates across sites.[18] If different parts of mtDNA mutate at different rates, then the application of a single mutation rate could seriously bias results. In particular, Wills looked at mutation rates for two types of mutations—transitions and transversions. A transition occurs when there is a change in nucleotide from a purine (A or G) to another purine or from a pyrimidine (C or T) to another pyrimidine. There are four possible transitions:

> "A" mutates into "G"
> "G" mutates into "A"
> "C" mutates into "T"
> "T" mutates into "C"

A transversion occurs when there is a change from a purine to a pyrimidine or vice versa. There are eight possible transversions:

> "A" mutates into "C"
> "C" mutates into "A"
> "A" mutates into "T"
> "T" mutates into "A"
> "G" mutates into "C"
> "C" mutates into "G"
> "G" mutates into "T"
> "T" mutates into "G"

Wills noted that transition rates are much higher than transversion rates and that transversion rates are better suited for dating a common mtDNA ancestor. He found that there is an average of 3.32 transversions between human mtDNA sequences and an average of 30.48 transversions between humans and chimpanzees. These numbers imply that the common mitochondrial ancestor of humans and chimpanzees lived 30.48/3.32 = 9.18 times as long ago as the common mitochondrial ancestor of all humans. He then looked at the minimum (4 million years) and maximum (7.4 million years) estimates of the human-chimpanzee split obtained from molecular studies. Applying the 4 million-year estimate means that the common human mitochondrial ancestor lived 4/9.18 = 0.436 million years ago =

436,000 years ago. Applying the 7.4 million-year estimate means that this ancestor lived 7.4/9.18 = 0.806 million years = 806,000 years ago.

Wills also noted that there is statistical uncertainty with such estimates and that it is more meaningful to examine the 95% confidence intervals, a range that we are 95% confident includes the "true" date we are trying to estimate (this is similar to the type of confidence intervals for poll results given on the news, where a result is given with a "margin of error"). Using the 4 million-year calibration value, this interval is between 336,000 and 481,000 years ago. If the 7.4 million-year calibration value is used, this interval is between 622,000 and 889,000 years ago. Wills' main point is that "Eve" may have lived much earlier than suggested by the initial studies, perhaps early enough to include a multiregional interpretation of the data.

Having reviewed the first analyses and some of the debate regarding "Eve," we can now turn to a more comprehensive review of how genetics can help in our search for a common ancestor. As with the Eve debates, our main goal is determining *when* and *where* our ancestors lived.

4.2 COALESCENT THEORY—WHEN DID THE COMMON ANCESTOR LIVE?

The approach used by Cann et al. and Vigilant et al. to estimate the date of a common mitochondrial ancestor focuses only on the accumulation of mutations over time. More appropriate methods, incorporating both mutation and genetic drift, have been developed based on a growing field of population genetics known as coalescent theory.[19] The basic principles of coalescent theory can be explained by the simple example shown in Fig. 4.8. Coalescent models work backwards in time to look at genealogical relationships between genes. In Fig. 4.8, there are four mtDNA sequences (the method is not confined to haploid inheritance, but it is easier to use mtDNA for illustration given the above discussion). These sequences are labeled A, B, C, and D. The genealogy shows the relationship of these sequences; the labels CA1, CA2, and CA3, identifying common ancestors, indicate places where the sequences coalesce. I have also included an ancestor of CA3 (CA4).

As we peer back in time, we see that C and D share an ancestor (CA1), who in turn shares an ancestor with B (CA2), who in turn shares an ancestor with A (CA3). In this case, the ancestor labeled "CA3" is a common ancestor of A, B, C, and D. Because CA4 is an ancestor of CA3, it is clear that both CA3 and CA4 are common ancestors of A, B, C, and D (the ancestor of a common ancestor is also a common ancestor). A basic principle of coalescent theory is that all genes will eventually coalesce to a single common ancestor. We are interested in making inferences about the *most recent common ancestor* (MRCA), which in this case would be CA3.

Coalescent theory shows that there will be a single most recent common ancestor when dealing with a trait such as mitochondrial DNA. Figure 4.8 starts with four DNA sequences. Earlier in time, when C and D have coalesced, there are only three sequences. Still earlier in time there are only two, and finally only one (CA3). Coalescent theory deals with the probability that any two sequences will coalesce in any given generation, which allows a prediction of how long, on average, it takes sequences to coalesce. This probability is related to population size—

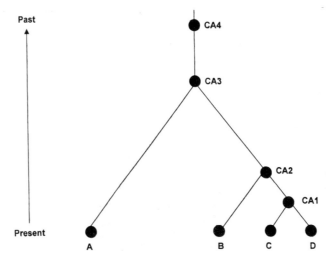

Figure 4.8. *Example of the principles of coalescent theory. The mtDNA sequences C and D have a common ancestor CA1, who in turn shares common ancestor CA2 with mtDNA sequence B. In turn, these individuals share a common ancestor CA3 with mtDNA sequence A. Ancestor CA3 is thus the common ancestor of mtDNA sequences A, B, C, and D. Another way of expressing this is to say that mtDNA sequences A–D coalesced at common ancestor CA3, which is the most recent common ancestor (MRCA). Note that the MRCA also has an ancestor (CA4 in this case), but we cannot make any inferences about this individual sequence.*

coalescence occurs more quickly in small populations because two individuals are more likely to be related in a small population than in a large population. Coalescent theory also considers the probability of mutations in any branch of the gene tree.

Much of the interest in application of coalescent theory is to estimate the age of the most recent common ancestor (CA3 in Fig. 4.8). For mitochondrial DNA, the average time until coalescence to the most recent common ancestor is approximately equal to $2F$ generations, where F is the female population size. If, for example, the total population consisted of 1000 females of reproductive age, and this population size remained constant, then the most recent common ancestor is estimated to have lived 2(1000) = 2000 generations in the past. Keep in mind that such estimates are expected averages; because of drift, any particular DNA sequence might coalesce much sooner, or later, than expected *on average.*

A similar estimate applies when we are looking at nuclear DNA, where the average time until coalescence is $4N$ generations, where N refers to the total number of reproducing adults, male and female. The reason for this difference (4 rather than 2) is that nuclear DNA is inherited from two parents, so we have to double the time until coalescence. Furthermore, we have to count both males and females in the total population size. Comparing these two estimates means that we expect that the average time until coalescence for nuclear genes is four times that for mitochondrial DNA. This difference is clear when considering that $4N$ is equal to $4(M + F)$ where M is the population size of males and F is the population size of females. If we assume that there are equal numbers of males and females ($M = F$) then the total population size is $N = M + F = 2F$ and the time until coalescence for nuclear genes is $4N = 4(2F) = 8F$, which is four times that for mitochondrial DNA ($2F$).

The relationship between time until coalescence (the date for the most recent common ancestor) and population size means that if we have an estimate of one of these quantities we can estimate the other one. If we have an estimate of the number of females, we can estimate the time until coalescence. For example, if a species had a population size of 3000 females, then the expected average time until coalescence for mitochondrial DNA would be $2F = 2(3000) = 6000$ generations. Likewise, if we have an estimate of the date of coalescence, then we could estimate the number of females. For example, if all of the mitochondrial DNA coalesced 8000 generations ago, then we could estimate the population size of females as $F = 8000/2 = 4000$ women.

To estimate one of these numbers we need to have an estimate of the other one. There are several approaches to solving this problem using coalescent theory. Several methods have been developed to estimate the average time until coalescence from data on DNA variation in living populations by computing the average number of DNA sequence differences between all pairs of individuals (known as the pairwise sequence divergence) and comparing this value with an expected value of DNA variation. These models estimate the average time until coalescence from measurements of genetic variation. However, the times are in mutational units rather than generations, so that we need to have an estimate of the mutation rate to translate the estimates into years. These methods also allow an estimate of the range of likely dates. Remember that the "true" date is likely to be less than or greater than our *average* estimate. Therefore, we need some way of estimating the probability that the "true" date lies within a given range. As discussed earlier, this range is usually given in terms of the 95% confidence interval, which gives us the range we are 95% confident includes the "true" date.

4.2.1 Estimates of Coalescent Dates from Mitochondrial DNA

Methods based on coalescent theory have been applied to the question of the age of "mitochondrial Eve." Alan Templeton reanalyzed some early data collected by Mark Stoneking and colleagues.[20] Templeton obtained an estimate of the time until coalescence in terms of mutational change and calibrated this using the 2% divergence rate used by Cann et al. This method produced an estimate of 290,000 years, similar to the date estimated by Cann et al. using a molecular clock. However, Templeton noted correctly that this is an expected average time until coalescence and the actual date could be smaller or larger because of the variance introduced through genetic drift. In this case, the 95% confidence interval (CI) was 152,000–473,000 years. This large range (over 300,000 years) is typical; coalescent theory tells us that there is a large likely range because of the variation in the impact of genetic drift over time.

Templeton also examined coalescence dates based on the data from the Vigilant et al. study, where they had used different dates for the split of chimpanzee and human lines to calibrate their mutation rate. Using their four million-year estimate, Templeton estimated the time until coalescence at 170,000 years with a 95% confidence interval of 102,000–256,000 years. When Templeton used their six million-year calibration, he estimated coalescence at 256,000 years with a confidence interval of 153,000–389,000 years. Templeton also noted that the upper limit could be even higher, perhaps as much as 800,000 years or more, depending on the exact date used for calibration.

In another analysis, Maryellen Ruvolo and colleagues used a coalescent model to estimate the date of the most recent common ancestor based on mtDNA sequences for the slowly evolving cytochrome oxidase subunit II gene.[21] Comparing human and chimpanzee sequences, they obtained a date of 298,000 years, with a 95% confidence interval of 129,000– 536,000 years, using a six-million year calibration for the split between humans and chimps.

Any attempt to reconstruct history from genetics must deal with potential confounding influences that might obscure the underlying patterns of relationship. Key among these is natural selection. If two populations are biologically similar, we must distinguish between a model whereby their similarity reflects common ancestry and a model where similarity reflects convergence due to natural selection. When using mitochondrial DNA to estimate dates of the most recent common ancestor, we also must take into account the possibility that the amount of genetic diversity (which is used to estimate the date) has been affected by natural selection.

It is often assumed that mitochondrial DNA (or at least certain sections of it) is *neutral*; that is, not affected by natural selection. Under this assumption, it doesn't matter whether you have one sequence or another, and genetic differences would not affect survival and reproduction. The genetic difference between any two individuals will then be solely determined by mutation and genetic drift, allowing us to use mathematical models based on mutation and drift to estimate the time since they shared a recent common ancestor. If the assumption of neutrality is incorrect, then obviously any attempt to link genetic similarity with estimates of common ancestry is potentially flawed. Although there is evidence of some selection for certain sections of mitochondrial DNA,[22] there is still debate over how much of an effect past selection has on our estimates of dates. If harmful mutations arose in mitochondrial DNA that were subsequently removed because of natural selection, then measures of diversity in living populations would be less than expected if all mutations were neutral that accumulated over time. This means that, under natural selection, our estimates of the dates of the most recent common ancestor are underestimates, and the true date could be much older.

Another potential problem is the uncertainty about the mode of inheritance of mitochondrial DNA. Assuming maternal inheritance, everyone has only one mitochondrial DNA ancestor in any given generation, and there is no recombination to confuse interpretation. How accurate is this assumption? There have been several studies suggesting the possibility of some significant *male* contribution. If true, then many of the results based on mitochondrial DNA may be flawed. A recent study by Phillip Awadalla and colleagues suggested evidence for a paternal contribution and recombination,[23] but additional analyses to date have suggested errors with the data, methods of analysis, and interpretation.[24] The assumption of strict maternal inheritance appears justified.

4.2.2 Estimates of Coalescent Dates from Other Genes

Although mitochondrial DNA has been widely studied, any answer to the question of modern human origins cannot rest solely on the results from a single trait. Because of the variation imposed by the evolutionary process (drift), we cannot rely on any single trait but must instead look at values for as wide a range of genetic traits as possible. It seems to me that a common problem with the invention of a new method of genetic analysis is the temptation to read too much into it (I am not assigning blame—I have been equally guilty). The flurry of research surrounding mitochondrial DNA reminds me of studies of blood groups earlier in the twentieth century. A number of these early studies routinely looked at genetic variation of a single trait (most often the ABO blood group) or a small number of traits. Because of variation due to drift and natural selection, any conclusions based on a single trait could be incorrect. On the other hand, they may indeed lie close to the true underlying pattern of population history. The best we can do with a single trait is to label results as suggestive but not conclusive and then to look at many other traits to see whether the same average patterns result. To estimate the date of the most recent common ancestor, we need to sample as many different traits as possible.

One obvious genetic trait is the Y chromosome. One of the 23 pairs of chromosomes in human beings is the sex chromosome pair. There are two different forms of sex chromosomes, X and Y. In most cases, a human female has two X chromosomes and a human male has one X and one Y chromosome. Because a female will always pass on an X chromosome to her offspring, the sex of any couple's offspring is determined by whether or not the male passes on the X or Y chromosome. Because the only way a male child results is from inheriting the Y chromosome from his father, the Y chromosome is paternally inherited. Most of the Y chromosome does not recombine. Analyses of genetic variation of the Y chromosome provide an interesting contrast to maternally inherited mitochondrial DNA, providing a possible glimpse of a genetic "Adam" to go along with "Eve."

Several studies have applied coalescent theory to Y-chromosome variation. Robert Dorit and colleagues examined a 729-base pair sequence of the Y chromosome and found no sequence variation among 38 males from across the world.[25] They argued that this lack of genetic variation could reflect recent natural selection, a small ancestral population, or a recent origin for the human species. When they applied a coalescent model to their data they estimated a date of 270,000 years, very similar to that obtained from mitochondrial DNA studies. The 95% confidence intervals (95% CI), however, were quite large however, ranging from 0 to 800,000 years!

Some other studies give dates similar to those based on mitochondrial DNA. In one study of Y chromosome diversity, Michael Hammer estimated a coalescence date of 188,000 years (95% CI = 51,000–411,000 years).[26] Hammer and colleagues later analyzed a larger and more comprehensive data set and obtained similar results—a coalescence date of 147,000 years and a 95% confidence interval of 68,000–258,000 years.[27] Peter Underhill and colleagues identified 22 different polymorphisms on the Y chromosome and got two similar sets of dates: 162,000 years (95% CI = 69,000–316,000 years) and 186,000 years (77,000–372,000 years).[28] More recently, Russell Thomson and colleagues analyzed the Y chromosome DNA sequences and obtained a more recent date of 59,000 years ago, with a 95% confidence interval of 40,000–140,000 years ago.[29] This value was based on a model that incorporated population growth (as suggested by the data). Even so, the confidence intervals overlap those from earlier studies and in general are consistent with the mitochondrial DNA analyses in suggesting a fairly young most recent common ancestor of living humans.

Coalescent theory has also been applied to diploid traits (those inherited from both parents). In general, the coalescent dates for diploid genes are larger than those for mitochondrial DNA and the Y chromosome. This difference is expected, because there are more ancestors at any given point in time for diploid genes. Several studies have analyzed genetic variation of genes located on the X chromosome. In one study, Wei Huang and colleagues analyzed the *ZFX* gene and estimated a date of 306,000 years with a 95% confidence interval of 162,000–952,000 years.[30] Henrik Kaessmann and colleagues analyzed a section of noncoding DNA on the X chromosome and estimated a date of 675,000 years (95% CI = 525,000–975,000 years).[31] Eugene Harris and Jody Hey analyzed the *PDHA1* gene and estimated that the most recent common ancestor lived 1.73 million years ago (95% CI = 1.35–2.16 million years).[32]

Other studies of nuclear DNA also suggest deeper ancestry than found for mitochondrial DNA. Rosalind Harding and colleagues investigated genetic variation in section of the beta-globin gene, estimating the age of the most recent common ancestor as 750,000 years ago with a 95% confidence interval of 400,000–1,300,000 years.[33] This study is particularly noteworthy because of evidence of ancient Asian ancestors as well as African ancestors. Several other studies of nuclear DNA also found suggestions of coalescent dates ranging back over a million years.[34]

4.2.3   Coalescent Dates and Population Size

The finding that some traits (especially mitochondrial DNA) provide estimates of roughly 200,000 years or so continues to be cited as evidence supporting an African replacement model. Even given the wide confidence intervals, these dates appear much too recent to support a multiregional interpretation where dispersal out of Africa began close to two million years ago. Alan Templeton notes that this interpretation is correct only if we make the assumption that the gene trees reflect population trees. If this assumption is not correct, and gene flow occurred between geographic regions, then a DNA sequence can spread geographically *at any time*. Templeton considers a date of 200,000 years compatible with both replacement and multiregional models, and the date itself cannot distinguish between these models. The dates tell us about the history of particular genes, which may not be an accurate reflection of the history of populations.

Alan Rogers and Lynn Jorde point out another problem.[35] As noted earlier, the expected average date of coalescence is directly proportional to population size. For mitochondrial DNA, we expect that the average time until the most recent common ancestor is $2F$ generations, or twice the number of adult females in the population. For nuclear DNA, the expected average time until the most recent common ancestor is $4N$ generations, or four times the number of adults (male and female) in the population. Therefore, coalescence dates will be higher in a large species and lower in a small species. Perhaps all we are seeing here is the genetic signature of past population size. Rogers and Jorde argue that this is indeed the case and that the date by itself is of little importance. The critical finding is the estimated species size; they argue that the 200,000 year date simply shows us that the species was too small to have been spread out over several continents. Their argument is discussed in detail in Chapter 7. For the moment, we need only consider the implication that coalescent dates *by themselves* do not necessarily resolve the modern human origins debate.

4.3 GEOGRAPHY AND GENE TREES—WHERE DID OUR ANCESTORS LIVE?

Although studies of nuclear DNA tend to give coalescent dates close to a million years ago, the dates from mitochondrial DNA and Y chromosome data generally provide estimates closer to 200,000 years ago (although frequently with very large confidence intervals). Although the older dates are compatible with an ancient origin of humans, the younger dates appear at first glance to support a recent African origin and replacement model. Maryellen Ruvolo argues that we should not focus on any specific gene but rather look at the distribution of coalescent dates across many genes and observe where they tend to cluster.[36] In her view, there is a clustering of dates much closer to 200,000 years ago, as required under the replacement model, than to the one to two million-year date presumed to be required under a multiregional model. Others, such as Templeton and Rogers and Jorde, argue that these dates are not directly relevant. We also need to consider the other side of the question raised by the initial analysis of "mitochondrial Eve"—*where* did our ancestors live?

Many genetic studies look at geographic patterns of genetic variation, attempting to infer from these patterns an historical perspective on a gene tree. As discussed above, the initial "Eve" studies produced a reconstructed tree that had two deep branches. One of these branches contained people with African ancestry, whereas the other contained people of all different ancestries, African and non-African. Apart from the problems in reconstructing such trees (and the result that some trees have an African root and some don't), Templeton has shown that the observed pattern of mitochondrial DNA variation can also be explained by a pattern of recurrent gene flow and isolation by distance throughout much of the Old World. In this sense, the tree is not a tree of population fissions but instead results from gene flow and genetic drift in a set of interconnected populations.

Templeton's approach has also been applied to Y chromosome variation, suggesting a complex pattern of past evolution. Hammer and colleagues used Templeton's methods to investigate geographic associations for 10 Y chromosome variants in 1544 men in 35 populations across the world.[37] This analysis indicates that the most common Y chromosome type originated in Africa and dispersed out of Africa replacing Y chromosome types throughout the Old World. However, there is also genetic evidence for subsequent gene flow into Africa from Asia. This result suggests that there has been genetic exchange between Africa and Asia throughout the time period back until coalescence (the past 150,000 years). The data are not compatible with a total replacement within the past 100,000 years, as suggested by some advocates of the replacement model. The evidence for intercontinental gene flow predates the supposed replacement.

Hammer and colleagues compared Y chromosome and mitochondrial DNA data and noted that although the coalescent dates are similar, the geographic history of each trait is somewhat different. The Y chromosome data suggest an initial African replacement of Y chromosome types in Europe and Asia *and* subsequent gene flow from Asia to Africa. The geographic pattern of mitochondrial DNA shows a different pattern characterized by short-range gene flow throughout the Old World during the past 200,000 years. One provocative suggestion is that this difference reflects sex differences in migration. The observed geographic associations are compatible with males moving more frequently over large distances, whereas females tend to move more frequently over shorter distances. Although this is suggestive, we must remember that only two traits have been examined.

Nuclear DNA has also been used to address the question of geographic origins. Harris and Hey looked at the geographic distribution of 10 different forms of the *PDHA1* gene on the X chromosome.[38] Their analysis revealed that seven of these forms occurred only in Africans and the remaining three occurred only in non-Africans. Furthermore, their gene tree showed two deep branches, one containing Africans and the other containing both Africans and non-Africans. This is the same overall pattern as found from mitochondrial DNA trees but with an estimated recent common ancestor dating to 1.73 million years ago, a date that is close to the emergence of the species *Homo erectus*. Further inspection of their gene tree showed that separation of African and non-African populations probably took place roughly 200,000 years ago. They concluded that population separation had begun before the origin of anatomically modern humans, although it is not clear whether a wide-scale multiregional interpretation is appropriate. Harris and Hey note that it is possible that the genetic evidence for population separation reflects events within a relatively small geographic region, perhaps confined to Africa.

Rosalind Harding and colleagues have provided the strongest genetic evidence for an ancient non-African genetic contribution to modern humans.[39] Their analysis focused on 16 haplotypes of a portion of the beta-globin gene found in 326 DNA sequences from across the world. As noted above, coalescent models estimated the age of the most recent common ancestor at roughly 800,000 years. They reconstructed a gene tree that shows the most likely past history of mutations that would give rise to these 16 haplotypes. Over half of the mutation events were estimated to have occurred more than 200,000 years ago, including several that are specific to Asian populations. The estimated age and geographic distribution of these Asian haplotypes "suggest that the ancestral human population was located in Asia, as well as in Africa, >200,000 years ago."[40] Furthermore, several haplotypes have a worldwide distribution that suggests gene flow back and forth between Africa and Asia in the past. This finding contrasts with a replacement model where the gene flow is in one direction—from Africa to Asia. If nothing else, such studies show us that the use of simple models, assuming a population "split" followed by isolation, are invalid.

4.4 SUMMARY

Genetic variation in living human populations can provide us with valuable insight into the past evolutionary history of our species. A primary goal of such work is to determine how geographically widespread our ancestors were over the past 200,000 years or so. If an African replacement model is correct, then all of our ancestors 200,000 years ago lived in Africa. Other human populations living outside of Africa became extinct. If, on the other hand, a multiregional model is correct, then our ancestors lived in more than one region; some lived in Africa and some lived elsewhere (although perhaps not *everywhere* throughout the Old World).

Where and when did our recent common ancestors live? Phylogenetic trees and coalescent theory provide ways of answering these questions, although the interpretation of the results is, as we have seen, not always as clear as we might think. Although much of the work performed to date has focused on mitochondrial DNA, an increasing number of studies focusing on other genetic traits are expanding our knowledge of human genetic variation and population history.

The mitochondrial DNA evidence is generally compatible with the African replacement model. This model predicts that the most recent common mitochondrial DNA ancestor lived in Africa roughly 200,000 years ago. Although some of the earlier analyses were flawed, there have been other studies confirming their basic finding of agreement with the replacement model. A basic principle of the scientific method is to compare observed reality with results expected under a given hypothesis. The fact that the mitochondrial DNA evidence generally fits the expectations of the African replacement model can be taken as strong support for that model. However, the match of reality with theoretical expectation does not constitute proof unless the data do *not* fit alternative hypotheses.

The problem here is that strong arguments can be made for a multiregional interpretation of the data. Consider first the African location of the common ancestor. Alan Templeton notes that although this is compatible with the African replacement model, it does not prove it. After all, the common mitochondrial ancestor had to live *somewhere*. Given genetic evidence that Africa had a larger population than other regions in the past (to be discussed in Chapter 5), it is not surprising to expect "Eve" to have lived in Africa. Templeton also notes a basic problem with the use of geographic origin as a test of the modern human origins hypotheses. If we had found an *non-African* common ancestor, then that observation would have ruled out the African replacement model. The reverse is not true; finding an African common ancestor does not necessarily rule out a multiregional model. Templeton's analysis of geographic variation, though not accepted by all, suggests a pattern of gene flow into and out of Africa over the past 200,000 years. In particular, his work suggests gene flow between Africa and much of Asia but a more recent expansion (and possibly replacement?) in Europe. I will consider the implications of this finding momentarily.

The date of the recent common mitochondrial ancestor is also compatible with the African replacement model. To some, this recent date (even given the large confidence intervals) points to a common ancestor that lived much too recently to be compatible with the initial dispersion of *Homo erectus* out of Africa into the rest of the Old World. According to some experts, any date more recent than a million years or more rejects a multiregional model. Again, the situation is not that clear, because the relationship between coalescent dates and population history can be complex and explained in different ways. Templeton argues that a common ancestor could arise at any time under a multiregional model involving gene flow between geographic regions. Rogers and Jorde also dispute the idea that the date tells us anything about the timing of the origin of a given population. They note that coalescence dates only tell us about the *population size* of our ancestors. They argue for a recent African origin using this different interpretation—if the size of our species was very small, then our ancestors could not have been spread out over several continents. In this context, the date *is* important, but only in terms of indirect evidence about population size.

Analysis from other traits is sometimes in agreement with mitochondrial DNA but sometimes is different. The coalescent dates for Y chromosome DNA are similar to those from mitochondrial DNA, whereas the dates from diploid nuclear DNA are much larger. The difference in dates is expected; on average, the coalescent dates for nuclear DNA is expected to be four times that of mitochondrial DNA or Y chromosome DNA. According to proponents of the African replacement model, the later dates are not that relevant; the important finding is a *recent* common ancestor for mitochondrial DNA and Y chromosome DNA. If, they argue, the multiregional model is correct, then these dates should be much larger—somewhere before one million years for mitochondrial and Y chromosome DNA and roughly four million years for other traits.

Of course, there are other questions to consider with coalescent dates, including accuracy in estimating mutation rates, statistical variation across genes, and the possibility of natural selection biasing results. Ultimately, even if all of these questions are satisfactorily answered, we are still left with a basic interpretive difference—what do these dates mean? Are they a reflection of the (recent) origin of our species, the population size of our ancestors, or both?

We must also consider additional findings relating gene trees to geography. Michael Hammer's studies of the Y chromosome support gene flow back to Africa from Asia within the past 200,000 years or so. Even if our species arose as a new species before that time, the pattern of evolution does not appear to be a simple case of continued expansion out of Africa. Rosalind Harding's study of beta-globin DNA also argues for ancient Asian genetic input to our species, dating back 200,000 years or more.

Putting these pieces together suggests that, regardless of what model is correct, Africa was an important region in recent human evolution (and, as I argue over the course of the following chapters, this importance is probably related to population size and geographic centrality). There is also evidence for an ancient Asian influence as revealed from Templeton's geographic analysis of mitochondrial DNA, Hammer's gene tree for Y chromosome DNA, and Harding's gene tree for beta-globin DNA. These findings suggest that our ancestors lived in at least two geographic regions—Africa and Asia.

What about Europe? On the basis of the findings presented in this chapter, the question of Europe is more complex. The Y chromosome and beta-globin analyses do not provide any evidence for an ancient European influence. Templeton's mitochondrial DNA analysis also suggests that Europe might represent a case of population replacement. One could argue for a model in which parts of Africa and Asia were involved with the transition from archaic to modern humans but replacement was the mode in Europe. If this is true, then we are seeing a mixture of evolutionary events, continuity in some regions and replacement in others. Although the European connection is still debated (the "fate" of the Neandertals), the evidence reviewed in this chapter suggests that a model of *total* replacement is not correct. We need to consider additional genetic evidence over the next four chapters before examining this possibility further.

4.5 POSTSCRIPT

Many of the mtDNA studies reported in this chapter have focused analysis on the control region of mtDNA, which represents only a fraction of the total mtDNA genome. A study by Max Ingman and colleagues was published in December 2000 while this book was in production, and deals with a global analysis of the entire mtDNA genome.[41] Many of the results confirm those based on the control region, including a relatively recent date for the most recent common ancestor of 171,500 years ago, an African root, and higher African diversity (to be discussed in the next chapter). While the authors interpret these data as evidence for a recent African origin of modern humans, the same alternative explanations discussed in this chapter still apply. To me, the increased details of the data, while important, have not solved the debate.

Chapter 5

Genetic Diversity and Recent Human Evolution

Diversity is a central concept in evolutionary biology. Diversity refers to differences between individuals within a species. Diversity is another name for variation, the amount of difference between one individual and another. Organisms vary. Look around at a group of people, and you will see both similarities and differences. Focusing for the moment on the differences, you will observe that some people are taller than others are and some people are darker than others are. Some have light hair, and some have dark hair. Look more closely, and you will see variation in terms of the size and shape of the skull, size of the nose, and other physical traits. If you look closely, you will see variation even within what you might consider a homogeneous group. Contrary to the old saying, they (whoever "they" are) do *not* all look alike. You can see variations within any group of humans, including within families. You probably are rather similar to any siblings you might have, but you are not identical.

Diversity exists at the genetic level for a variety of characteristics that we cannot detect visually. Some people have different blood groups than others or different blood proteins or enzymes. For many traits, we see diversity in the genotypes and phenotypes typically studied by anthropologists. Delve even further, and we can see diversity at the molecular level, with different sequences of DNA.

Diversity is necessary for evolution. Without variation for natural selection to act on, a species cannot adapt to new environments and is likely to become extinct. We also see the importance of diversity in biological conservation, where species with low diversity and/or numbers are considered in potential danger. One famous example is the cheetah, a large member of the cat family found in Africa, which shows very low levels of genetic variation.[1] Low diversity can lead to a number of problems, including failure to adapt to changing environments, increased mortality, greater disease susceptibility, and lowered fertility.

5.1   GENETIC DIVERSITY AND EVOLUTION

The level of genetic diversity in a population provides clues as to past events. All of the evolutionary forces (mutation, natural selection, genetic drift, gene flow) affect levels of genetic variation within populations. Mutation introduces new genetic variants into a population and will increase diversity. The rate at which new variation is introduced is generally slow over a short time scale but can accumulate over the thousands of generations we deal with in the study of modern human origins. Natural selection can increase or decrease diversity within a population, depending on the specific type of selection in operation. If harmful alleles are being selected against, or helpful alleles are being selected for, then natural selection lowers diversity, because one allele is being replaced by another. If, however, selection is for the heterozygotes, then diversity is likely to be increased.[2] Genetic drift acts to reduce diversity in a population. Given enough time, the eventual fate of any allele is to be completely fixed or eliminated from a population. This means that, over time, allele frequencies will tend to move toward fixation (a frequency of 1.0) or extinction (a frequency of 0.0). The replacement of one allele by another results in a loss of diversity (because there is no diversity if an allele has a frequency of 1.0—it is the only one in the population's gene pool). Gene flow can increase diversity within a population by introducing new alleles from other populations.

If we focus on neutral traits (those not affected by natural selection), then the level of genetic diversity within a population results from a balance between mutation, genetic drift, and gene flow. Mutation and gene flow increase diversity, whereas drift decreases diversity. Mathematically, within-population diversity will increase at first because of mutation and gene flow. Over time, genetic drift will operate in the opposite direction by decreasing diversity. Eventually, an equilibrium will be reached whereby the increase in diversity of each generation due to mutation and gene flow is countered by the decrease in diversity due to genetic drift. A simple example of what this process looks like is shown in Fig. 5.1, which represents the change in diversity over 500 generations.[3] Diversity increases quickly for the first

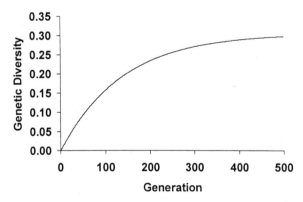

Figure 5.1. Changes in genetic diversity within a population over time. The simulated population starts with zero diversity and has a constant population size of 100 reproductive adults, a migration rate of 0.001 per generation, and a mutation rate of 0.0001. See chapter notes for the computational formula.

200 generations. The rate of increase in diversity slows down, and the population is near equilibrium after roughly 400 generations.

A simple analogy helps illustrate this process. Take an empty paper cup and punch a small hole in the bottom with a pencil (I got good results with a hole about a quarter of an inch wide). Hold the cup under a water faucet and turn on the water to a slow trickle. Water will enter the cup from the faucet and exit the cup via the hole at the bottom. Increase the amount of water flow very slowly. After some experimentation with flow rate, you will see a situation in which the water level in the cup will rise at first and eventually reach a level in which the amount of water entering the cup is equal to the amount of water leaving the cup. The water entering the cup is analogous to the addition of new diversity through mutation and gene flow. The water leaving the cup through the hole in the bottom is analogous to diversity lost through genetic drift. The level of the water in the cup is analogous to the level of diversity within a population. When you have adjusted the water flow just right, the level will not change and is at equilibrium.

This analogy helps illustrate several other features of genetic diversity. If you turn the water up, the level of water will increase in the cup at a faster rate and reach a higher equilibrium level. This is analogous to an increase in the rate of mutation and/or gene flow, which will result in a higher equilibrium diversity that will be reached at a faster rate. If you make the hole in the cup larger, more water will be lost and the equilibrium level will drop. If the hole is big enough, the water will drain out very quickly. This is analogous to an increase in genetic drift, which would accompany a reduction in population size (because the expected effect of drift is inversely proportional to population size).

These simple analogies help illustrate the point that the rate of change in genetic diversity and the ultimate equilibrium level are a function of the balance between mutation, gene flow, and genetic drift. We can also make several other important inferences about changes in genetic diversity over time. Mutation leads to an increase in genetic diversity over time. The more time that has elapsed, the more mutations will have accumulated and the higher the level of genetic diversity. Assuming all other factors to be equal, differences in diversity in two populations may therefore reflect differences in the amount of time that has elapsed. Imagine two populations starting with zero diversity (there is only one allele). Over time, new alleles will be introduced through mutation. The more time that has passed, the greater the accumulation of mutation. Now, imagine further that these two populations both experience the same level of genetic drift and that there is no gene flow. Finally, imagine that one population is 500 generations old and the other population is 100 generations old. The older population will have had time to accumulate more mutations and will therefore have a greater level of genetic diversity than the younger population. Under this simple model, any difference in diversity will reflect the difference in a population's age. This means that, *under certain assumptions*, differences in diversity can be used to make inferences about differences in the age of the populations. As we shall see later, the critical issue is "under certain assumptions."

Differences in other variables can also lead to differences in genetic diversity. Imagine a somewhat different situation with two populations of the same age, but having different population sizes, such as 1000 for one population and 200 for the other population. If we ignore gene flow for the moment and assume that both populations start with zero diversity, we would expect each population to show an increase over time, leading to a balance between mutation and genetic drift. However, the expected effect of genetic drift will be greater in the smaller population. All other things being equal, larger populations will show greater diversity than small populations, who lose more diversity because of the greater effect of genetic drift. Under this simple model, any difference in genetic diversity will reflect the difference in population size. Under certain assumptions, differences in diversity can be used to make inferences about population size.

There are other possible models, including varying levels of gene flow, changes in population size over time, and other complicating factors. The point here is that conceptually different models can lead to the same result—a difference in genetic diversity. Which one of these (or others) is the best description of what has happened in the past? Do differences in diversity reflect differences in a population's age, differences in population size (and hence drift), or some other combination of factors? This question is pertinent to the modern human origins debate. As described below, many traits show the highest levels of genetic diversity in sub-Saharan African populations. Some researchers have suggested that higher African diversity reflects a greater age of African populations, as expected under the African replacement model. If modern humans arose first in Africa and then much later some of them dispersed into other regions, then the non-African populations would be younger. As such, the non-African populations would have had less time to accumulate mutations and would have lower levels of diversity, which is exactly the pattern that we observe today. Although observed reality is compatible with the African replacement model, this fit does not prove the model correct unless other viable alternatives are ruled out. From the examples given above, it should be obvious that there are other alternatives, particularly a model in which Africa had a larger population size than other regions. The observed pattern of greater African diversity might reflect differences in population size, which could fit either African replacement or multiregional models. How can we distinguish between these alternatives? I will return to this question after a brief discussion of the actual evidence on regional differences in genetic diversity.

5.2 MEASURING GENETIC DIVERSITY

How is genetic diversity measured? What exactly does this term mean? To examine the relationship between the geographic distribution of human genetic diversity and the question of modern human origins, we need to consider some of the different ways in which diversity is defined and measured. Although genetic diversity has differing definitions depending on the specific type of trait being looked at, the basic principles relating diversity to evolutionary processes apply to all.

One of the simplest measures of diversity is heterozygosity, which is the expected proportion of individuals in a population that are heterozygous (having two different alleles, one from their mother and a different one from their father) rather than homozygous (having two copies of the same allele).[4] As an example, consider the MN blood group discussed in Chapter 2. There are two alleles—*M* and *N*. With only two alleles, everyone has one of three different genotypes—*MM*, *MN*, or *NN*. If you have two copies of the same allele (the genotype *MM* or *NN*), you are homozygous for this locus. If you have one of each allele (the genotype *MN*), you are heterozygous. Heterozygosity is the proportion of individuals expected to have the genotype *MN*.

The level of heterozygosity in a population is computed based on the frequencies of the two alleles. If we use the symbol p to refer to the allele frequency of M, the heterozygosity is equal to $2p(1 - p)$. Imagine a population in which the frequency of M is equal to 0.6. Because there are only two alleles, the frequency of the N allele is obviously $1 - p = 1 - 0.6 = 0.4$. The heterozygosity for the *MN* blood group in this population is $2(0.6)(0.4) = 0.48$. Now, consider another population in which the frequency of the M allele is equal to 0.7. The frequency of the N allele is $1 - 0.7 = 0.3$, and the heterozygosity is equal to $2(0.7)(0.3) = 0.42$. Comparing the two heterozygosities, we see that the first population has more genetic diversity for this trait (0.48) than the second population (0.42). Note that if there is only one allele present in a population (say $p = 1$ or $p = 0$), then there is zero diversity, because $2(1)(0) = 0$. This makes intuitive sense, because if everyone has two copies of the same allele, then everyone is genetically the same.

Computing heterozygosity is a little different if there are more than two alleles. The method used here is to square all of the allele frequencies, add them up, and subtract the total from 1. As an example, consider the alpha haptoglobin protein gene located on chromosome number 16. There are three alleles found in many human populations: HPA*1S, HPA*1F, and HPA*2. One study[5] found the following allele frequencies for a population in Nigeria:

$$HPA*1S = 0.258$$

$$HPA*1F = 0.473$$

$$HPA*2 = 0.269$$

To compute the heterozygosity for this population, we compute the sum of the squared allele frequencies, which is equal to

$$(0.258)^2 + (0.473)^2 + (0.269)^2 = 0.0666 + 0.2237 + 0.0724 = 0.3627$$

We then subtract this sum from the number 1 to get

$$\text{heterozygosity} = 1 - 0.3627 = 0.6373$$

This method provides us with the heterozygosity for a single locus. In any actual study, we would want to obtain an overall estimate of heterozygosity, which would be computed as the average over all loci.[6] This method of computing heterozygosity is used when the data consist of allele frequencies.

There are a number of different measures of diversity that are used when dealing with DNA sequences depending on whether we are looking directly at nucleotide sequences, restriction-site differences, or other measures of DNA variation.[7] One commonly used measure is the average number of nucleotide differences per site in a population. Consider the following four hypothetical DNA sequences, each representing nine base pairs. The numbers along the top (1 to 9) are labels for the nine nucleotide sites.

	1 2 3 4 5 6 7 8 9
sequence #1	A A C C T G A G C
sequence #2	A A G C T G A G C
sequence #3	A A C C A G A G C
sequence #4	A A C T T G A G C

The basic method involves counting the number of differences between all pairs of sequences. Thus we have to compare sequence #1 with sequences #2, #3, and #4. Then we compare sequence #2 with sequences #3 and #4. Finally, we compare sequence #3 and sequence #4. All possible pairs are counted. In this hypothetical example there are four sequences and six different comparisons: 1-2, 1-3, 1-4, 2-3, 2-4, 3-4. If there were 5 sequences, there would be 10 comparisons. If there were 6 sequences there would be 15 comparisons. The general rule is that if there are n sequences, there will be $n(n-1)/2$ comparisons.

Start by comparing sequence #1 and sequence #2. There is only one difference, which occurs at the third position—sequence #1 contains C whereas sequence #2 contains G. We next compare sequence # with sequence #2; there is also one difference, this time for the fifth position. We then continue to compare all different pairs. We end up with the following list of comparisons:

sequence #1 and sequence #2	1 difference
sequence #1 and sequence #3	1 difference
sequence #1 and sequence #4	1 difference
sequence #2 and sequence #3	2 differences
sequence #2 and sequence #4	2 differences
sequence #3 and sequence #4	2 differences

The total number of differences observed from all pairwise comparisons is 9. We then divide this number by the number of comparisons (6), which gives 9/6 = 1.5 differences per comparison. To know the average number of differences per nucleotide site, we need to further divide this result by the number of analyzed sites per sequence (9). This gives the final result of 1.5/9 = 0.167. This number refers to the mean sequence divergence, which is the average number of substitutions per site; the higher this number, the greater the genetic diversity. Similar measures have been developed for restriction-site data.

Different measures are used to assess diversity in quantitative complex traits, such as anthropometrics and skin color. Except for some cases in which sufficient data exist to estimate the degree of genetic variation, most studies using quantitative traits focus on phenotypic variation—which is what is actually observed and measured. Variation in the phenotype of a quantitative trait, such as head length or skin color, is a function of both genetic and environmental variation. Much of the time we simply take phenotypic variation as proportional to genetic variation, which means we must make the assumption that heritability (the relative proportion of total variation due to genetic variation) is the same across populations. In some cases, this assumption appears to be valid, allowing us to make genetic inferences about diversity from phenotypic measures.

As will be familiar to anyone who has taken an introductory course in statistics, there are two basic statistics used to analyze quantitative traits—the mean and the variance. The mean is simply the average value of a trait in a sample. The following list is from one of my data sets—the numbers refer to the head length (measured in millimeters) for 15 women who lived in the town of Claddagh in County Galway, Ireland.

180	185	191
180	185	191
182	186	192
184	188	196
185	188	203

We can see from these numbers that there is diversity—not all women have the same head length. The values range from a low of 180 mm to a high of 203 mm. To compute the mean head length in this sample, we simply add up all the values and divide by the number of observations. The sum of all 15 numbers is $180 + 180 + 182 + \ldots + 203 = 2816$. We divide this number by 15 (the number of observations) to get a mean of $2816 / 15 = 187.7$ mm. We can express this process using a standard statistical formula

$$\bar{X} = \frac{\sum X}{n}$$

The symbol for the mean is $\bar{X}$. The Greek symbol sigma (Σ) is mathematical shorthand for "summing." In this case, the expression ΣX is shorthand for "add up all the individual values of variable X." The symbol n refers to the sample size, which is the number of observations.

Variance is a measure of how much the individual values differ from the mean. If everyone had the same value as the mean, then the variance would be equal to zero. We see, however, that there is some variation; although some of the women had values very close to the mean, others had values smaller or larger than the mean. The variance is an estimate of the average squared difference to the mean. In mathematical terms, the variance (V) is computed as

$$V = \frac{\sum (X - \bar{X})^2}{n - 1}$$

Computation is straightforward. Take the value of the first observation ($X = 180$), subtract the mean ($\bar{X} = 187.7$) from it, and then square it. This gives $(180 - 187.7)^2 = (-7.7)^2 = 59.29$. Do the same thing for each of the 15 observations and add them all up. This gives $\Sigma(X - \bar{X})^2 = 532.93$. Finally, this value is divided by $n - 1$, which gives a variance of

$$V = \frac{\sum (X - \overline{X})^2}{n-1} = \frac{532.93}{15-1} = \frac{532.93}{14} = 38.07$$

(If you are wondering why we divided by $n - 1$, rather than n, to get an estimate of the average spread around the mean, consult any statistics text for the long answer— the short answer is that this method gives us an unbiased estimate). Another measure of variation is the standard deviation, which is simply the square root of the variance, which here is equal to $\sqrt{38.07} = 6.17$. In either case, the higher the number, the greater the phenotypic diversity.

5.3 THE GEOGRAPHIC DISTRIBUTION OF HUMAN GENETIC DIVERSITY

For many traits, genetic diversity is greatest within sub-Saharan African populations and people with sub-Saharan African ancestry.[8] This pattern has been found in studies of mitochondrial DNA.[9] Figure 5.2 shows the results of four different studies of mitochondrial DNA in living human populations. The actual numbers vary across studies because of differences in the specific region of mtDNA analyzed and the use of somewhat different measures of diversity. Nonetheless, all four studies show that genetic diversity is greatest among those individuals with sub-Saharan African ancestry.

The same pattern has been found in studies of microsatellite DNA variation.[10] The results of four studies looking at regional differences in heterozygosity are shown in Fig. 5.3. These analyses were based on between 8 and 60 micro-satellite DNA loci. In each case, the African sample shows the highest level of heterozygosity.

Greater African diversity is also seen in quantitative traits.[11] Figure 5.4 shows the average within-region phenotypic variance based on 57 measurements of 1,159 crania. The highest level of variation is found in the sub-Saharan African sample. Figure 5.5 shows phenotypic variation in human skin color based on studies of 98 populations broken down into eight geographic regions. As with craniometrics, phenotypic variation is highest in sub-Saharan African populations.

Not all genetic data show the same pattern. In a study of *Alu* insertion polymorphisms, Mark Stoneking and colleagues found that while Africa showed very high diversity compared with much of the rest of the world, diversity in western Asia was actually somewhat higher.[12] Studies of classic genetic markers (blood groups and blood protein and enzyme markers) and nuclear restriction fragment length polymorphisms (RFLPs) usually show Europe to have the highest average heterozygosity. Why do these traits show a different pattern than those described above? One explanation is ascertainment bias; many of the classical genetic markers and RFLPs were first discovered in European samples. Samples from Europeans were used to detect genetic variants, meaning that, by definition, these traits will be highly variable within Europe. Alan Rogers and Lynn Jorde have shown how this bias will lead to higher levels of European diversity for such traits.[13] Another potential factor is mutation rate. I have shown

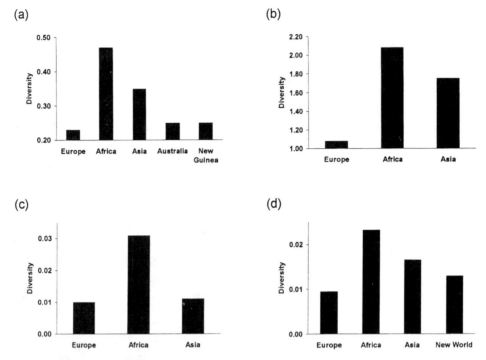

Figure 5.2. *Regional mtDNA diversity in four separate studies. a. Percentage sequence diversity esti-mated from RFLP analysis (Cann et al. 1987). b. Percentage sequence diversity for a 610-bp sequence of the mtDNA control sequence (Vigilant et al. 1991). c. Nucleotide diversity for a 200-bp sequence of mtDNA hypervariable sequence II (Jorde et al. 1995). d. Nucleotide diversity for the D-loop region (reported in Bowcock et al. 1994).*

elsewhere that the level of African diversity depends in part on mutation rate.[14] Traits with low mutation rates, such as classic genetic markers and RFLPs, will not show higher African diversity, whereas traits with higher mutation rates, such as mitochondrial DNA and microsatellite DNA, will show higher African diversity. It is likely that the combination of ascertainment bias and low mutation rate is respon-sible for the different patterns of variation seen in classic genetic markers and nuclear RFLPs.

An interesting geographic pattern of diversity within local populations was noted by Sarah Tishkoff and colleagues, who looked at microsatellite alleles for the CD4 locus on chromosome 12 in 42 local populations from across the world.[15] Sub-Saharan African populations showed higher levels of diversity in terms of both the number of haplotypes and heterozygosity. They also looked at an *Alu* deletion poly-morphism at the same locus characterized by the loss of 256 bp out of a 285-bp sequence. There were 12 different forms of the microsatellite repeat, which along with two types for the *Alu* deletion (present/absent), gave 24 possible combinations observed in human populations. The most interesting aspect of their study was the geographic pattern of these combinations. The *Alu* deletion occurred in a number of these combinations in sub-Saharan African populations but usually in only one combination outside of Africa. This nonrandom association led Tishkoff and col-

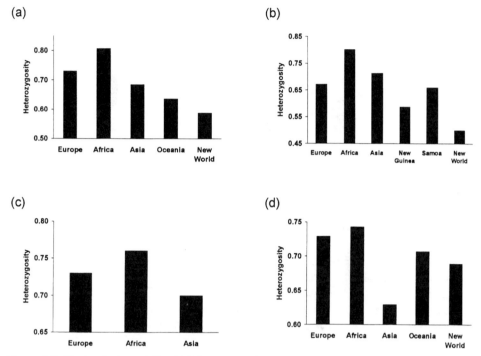

Figure 5.3. *Regional microsatellite DNA heterozygosity in four separate studies. a. Based on 30 loci (Bowcock et al. 1994). b. Based on 60 loci (Jorde et al. 1997). c. Based on 8 loci (Deka et al. 1995). d. Based on 20 loci (Pérez-Lezaun et al. 1997).*

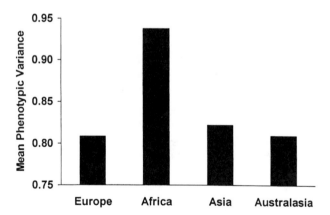

Figure 5.4. *Regional craniometric variation based on the average standardized variance for 57 craniometric variables (Relethford and Harpending 1994).*

leagues to suggest that non-African populations all shared a particular combination because of a single geographically restricted origin. All other combinations may have been lost in an initial bottleneck, perhaps associated with a recent origin of our species. They further argued that, given a constant mutation rate, this move-

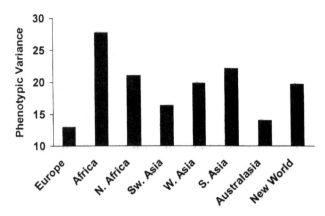

Figure 5.5. *Regional phenotypic variation in skin color (Relethford 2000a).*

ment from Africa took place about 100,000 years ago, consistent with the African replacement model. Other interpretations have been offered, including genetic drift outside of Africa because of smaller population size.[16] Some have questioned the extent to which any single trait can be used to reconstruct the history of populations (as opposed to the history of a particular gene)—a criticism also true of mitochondrial DNA.

5.4 EVOLUTIONARY INTERPRETATIONS OF AFRICAN GENETIC DIVERSITY

There is growing evidence that many traits show higher levels of genetic diversity in sub-Saharan African populations than in any other geographic region. Why? There are two possible explanations. The first is that we are seeing a genetic signature of population history in terms of population age—Africa shows greater diversity because it is older. The second possible explanation is that we are seeing a genetic signature of demographic history—African shows greater diversity because the population size was larger. The first of these interpretations is only compatible with the African replacement model, whereas the second is compatible with either African replacement or a multiregional model.

5.4.1 A Recent African Origin of Our Species?

Rebecca Cann and colleagues made the point in their 1987 *Nature* paper that higher African diversity for mitochondrial DNA is an expected consequence of an African replacement model. Their reasoning is that older populations will accumulate more mutations and will consequently show higher levels of diversity. Mark Stoneking and Rebecca Cann later stated this rationale succinctly: "If one accepts that mtDNA mutations are largely neutral, then their occurrence and accumulation are mostly a function of time; the more variability a population possesses, the older it is."[17]

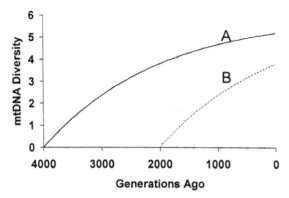

Figure 5.6. *Evolution of mitochondrial DNA diversity in two simulated populations, each consisting of 2000 reproductive females. The only difference between these two populations is that population A starts 4000 generations in the past, whereas population B starts 2000 generations in the past and has not had as much time to accumulate mutations and diversity. As a result, population A shows more diversity when measured in the "present" (generation 0). See chapter notes for computational details.*

Because Africa shows the highest level of within-group diversity, it is therefore the oldest regional population. This inference fits the African replacement model, which predicts that modern humans *first* arose in Africa and then *later* split into African and non-African lines.

The expected relationship between diversity and population age[18] is illustrated in Fig. 5.6. This graph represents the evolution of mitochondrial DNA diversity within two hypothetical populations, labeled A and B. Each population consists of 2000 females and remains constant in size over time. The only difference between these two populations is their age; population A is 4000 generations old, and population B is 2000 generations old. The expected values of mitochondrial DNA diversity were computed using a standard formula (see chapter notes). Because population size is the same for both populations, we expect that *eventually* both populations will have the same level of diversity at equilibrium. However, the graph shows only what has happened over the past 4000 generations. Population A began accumulating diversity 4000 generations ago, whereas population B began accumulating diversity 2000 generations ago. When viewed from the present day (0 generations ago), population B shows less diversity because it is not as old.

According to the African replacement model, our species arose first in Africa. A portion of the species later split off and left Africa, which in turn split further and dispersed into other geographic regions. The formation of a new daughter population splitting off from a parental population is a well-known pattern in evolutionary biology. *If* diversity in the daughter population is less than in the parent population, then present-day levels of diversity should reflect this past history. Because non-African populations have less genetic diversity, it seems reasonable to infer that they have not been around as long as the African parent population. This inference depends on the critical assumption that a newly formed daughter population will have less diversity than the parent population. In other words, diversity in the daughter population is "reset" to a low value. If this "reset" does not happen, then the daughter population will have the same diversity as the parent population, and we could not make any inferences about the relative age of different regional populations.

How can diversity be lower in a daughter population? The answer lies with population size and genetic drift. A common evolutionary pattern is for a daughter population to be smaller in number than the parent population. Picture, for example, a population of 50 people splitting off from a larger population of 5000 people. These two groups will henceforth show different levels of genetic drift, with the smaller population being affected more. The effect of genetic drift is to decrease variation. The level of diversity in a smaller daughter population will begin to decline to approach a new, and lower, equilibrium value.

To understand this process, consider my previous analogy of water dripping from a faucet into a paper cup with a hole in the bottom. If you widen the hole in the bottom, the water already in the cup will drain out faster. If the rate at which the water drips remains the same, then the wider hole at the bottom will result in the water level going down. Likewise, an increase in genetic drift brought about by a reduction in population size will result in a loss of diversity over time. A major change in population size is called a bottleneck and results in increased genetic drift.

Figure 5.7 shows an example of the effects of a bottleneck on mitochondrial DNA diversity (the same process can be modeled for other measures of diversity as well).[19] This model starts with a parent population of 5000 females. A smaller population of 50 females splits off. The graph shows the level of diversity within this daughter population after the initial bottleneck; the level drops to approach a new (and lower) equilibrium value over time.

There are two key factors affecting the decline in genetic diversity in a bottleneck—magnitude and duration. Magnitude refers to the size of the bottleneck, and duration refers to how long the bottleneck lasts. The influence of both factors is

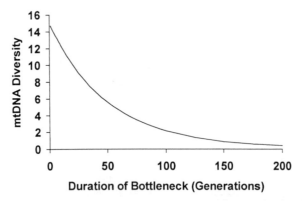

Figure 5.7. Simulation of the effects of a reduction in population size on mitochondrial DNA diversity. In this example, a daughter population of 50 females splits off from an initial parental population of 2000 females that had reached equilibrium. The smaller daughter population will lose diversity over time until it reaches a new equilibrium value based on its population size. This simulation shows two effects: 1. A reduction in population size will lead to a reduction in diversity. 2. It takes time for this to happen. See chapter notes for computational details.

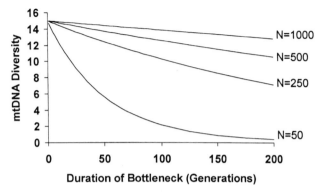

Figure 5.8. *Simulation of the effects of a reduction in population size on mitochondrial DNA diversity. This graph shows the same process described in Fig. 5.7 but for four different values of population size in the daughter population (N = 50, 250, 500, and 1000). In each case, the daughter population loses diversity over time as it approaches a new equilibrium value with less diversity. Note that the larger the size of the daughter population the slower this decline, such that there is very little reduction in diversity when the daughter population consists of 1000 females but considerably more when it consists of 50 females.*

shown in Fig. 5.8, which tracks four different daughter populations of varying size over time. The effect of bottleneck magnitude is clear—the smaller the daughter population, the greater the reduction in diversity. Figure 5.8 also shows the effect of bottleneck duration, where the longer the bottleneck, the greater the reduction in diversity.

In order for us to infer that diversity tracks population age, we must assume that the bottleneck that took place when some modern humans left Africa was severe and long enough to make a noticeable reduction in diversity. The scenario is further complicated by the fact that over time the daughter population would grow in size, thus countering the effect of the bottleneck to some extent. To illustrate the general principles, I performed a simple simulation[20] that started with a bottleneck occurring 5000 generations in the past when a daughter population of 50 split off from a parent population of 5000. I let the population remain at 50 females for 100 generations and then allowed it to grow instantaneously back to 5000 females for the remainder of the simulation (4900 generations). The sudden reduction and later sudden expansion is obviously unrealistic. but a number of studies have found it to approximate closely more complex patterns of demographic change, including more gradual reductions and increases in size. Figure 5.9 shows the effects of this changing demographic history on the expected level of mitochondrial DNA diversity in the daughter population. The initial bottleneck leads to a major reduction in mtDNA diversity very quickly (analogous to punching a larger hole in the bottom of the paper cup). When the population grows in size back to 5000 females, the loss of diversity is halted and then reversed (analogous to patching the hole in the bottom of the paper cup and allowing water to accumulate again). Very slowly, the population begins to gain back the diversity lost because of the bottleneck.

The simulation shown in Fig. 5.9 would seem to fit our observations of African and non-African diversity in that a reduction in population size leads to a notice-

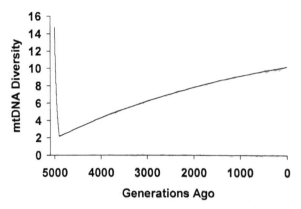

Figure 5.9. *Simulation of the effects of a reduction in population size, followed by a subsequent increase in population size, on mitochondrial DNA diversity. A daughter population of 50 females splits off from a parental population of 5000 females 5000 generations in the past. The daughter population remains at a size of 50 females for 100 generations, increases in size to 5000 females, and then remains at this size for the next 4900 generations. The initial reduction in diversity is caused by the drastic reduction in population size and occurs very quickly. Diversity begins to increase when population size increases, but at a slower rate. The difference in the rate of change before and after the recovery in population size is a function of population size—smaller populations will change more quickly. See chapter notes for computational details.*

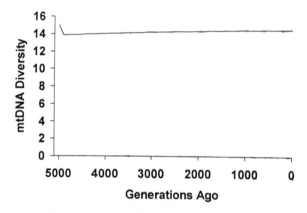

Figure 5.10. *Simulation of the effects of a reduction in population size, followed by a subsequent increase in population size, on mitochondrial DNA diversity. A daughter population of 1000 females splits off from a parental population of 5000 females 5000 generations in the past. The daughter population remains at a size of 1000 females for 100 generations, increases in size to 5000 females, and then remains at this size for the next 4900 generations. The initial reduction in diversity because of the bottleneck is minimal, particularly when compared to Fig. 5.9. See chapter notes for computational details.*

able reduction in diversity. However, this will not always be the case. The specific outcomes depend on the magnitude and severity of the bottleneck. Figure 5.10 shows the results of a similar simulation, keeping all conditions the same but using a bottleneck population size of 1000 rather than 50. There is a slight initial reduction in diversity accompanying the formation of a smaller group but not as much as in the previous example. Consequently, the difference in diversity between the daughter and parent populations is minimal.[21]

A bottleneck can result in a significant loss of genetic diversity under certain conditions: 1. The magnitude of the bottleneck is large (i.e., the daughter population is small); 2. The duration of the bottleneck is long; and 3. The amount of time after the bottleneck is not large enough to completely counter the impact of the bottleneck. We lack direct information on such history (indeed, this is part of the history we are trying to recover) and can only point out that a succession of population splitting *could* generate the geographic pattern of regional genetic diversity we observe today. However, a number of researchers suggest that a bottleneck sufficiently large enough and long enough to reset diversity is unlikely. Alan Rogers and Lynn Jorde investigated the half-life of convergence to a new equilibrium (the time it takes to reduce change the level of mtDNA diversity half of the way to a new equilibrium).[22] They show that it takes roughly $0.693N_f$ generations to reduce mitochondrial DNA diversity by half, where N_f is the number of reproductive females (similar arguments can be made for other traits). In order to "reset" diversity in a daughter population several half-lives must elapse. One half-life reduces diversity 50%, two half-lives reduce it 50% of the remaining, or 25%, giving a total of 75% reduction after two half-lives. Three half-lives results in a total reduction of $50\% + 25\% + 12.5\% = 87.5\%$, and four half-lives results in a total reduction of $50\% + 25\% + 12.5\% + 6.25\% = 93.75\%$. The time needed for four half-lives is approximately $4(0.693N_f) = 2.77N_f$. Thus it would take a bottleneck of 1000 females roughly 2770 generations to reduce the level of mtDNA diversity 94% to the new equilibrium. It seems unlikely that a bottleneck could last that long. If the bottleneck were more severe in magnitude, it would not have to last as long. For example, it would take 100 females 277 generations, it would take 50 females 139 generations, or it would take 10 females 28 generations to reduce mtDNA diversity 94% to the new equilibrium. At the lower end, it seems unlikely that so small a population would be able to survive that long. This does not mean that bottlenecks can't affect levels of genetic diversity—they do—but rather that they are unlikely to be so severe and/or long-lasting to reduce diversity enough to allow us to track the age of a population simply by comparing its level of diversity to other populations.

A further complication is the fact that a phylogenetic branching model, in which each population gives rise to other populations through splitting, is not useful for describing our species. The model assumes that there is no gene flow between groups after they split. In reality, humans have been quite mobile, so that it is not reasonable to use a model in which there is no migration between regions. Gene flow will counter the loss of diversity in groups, requiring us to assume an even more severe and/or long-lasting bottleneck to allow any inferences about a population's age from levels of diversity.

This does not mean that the African replacement model is incorrect; rather, it just means that regional differences in diversity cannot be taken as an index of relative population age. A number of researchers have suggested that a more appropriate way of explaining regional differences in diversity is regional differences in population size. As shown below, variation does not rule out an African replacement, but it also does not require one. Regional differences in population size are compatible with all models of modern human origins.

5.4.2   Regional Variation in Population Size?

Genetic diversity is a function of population size. All other things being equal, larger populations will be more genetically diverse than smaller populations. A simple example of the influence of population size is shown in Fig. 5.11. Here, I used the same methods to compute mitochondrial DNA diversity as in Fig. 5.6 but with different starting values. This graph shows two hypothetical populations, A and B, both 4000 generations old. The only difference between these two populations is their size—population A consists of 2000 females whereas population B consists of 1330 females. Both populations show the expected increase in diversity over time, but population B does not increase as fast because of its smaller size. Smaller population size results in greater genetic drift, which retards the increase in diversity due to mutation.

If we compare the results of the two simple simulations in Figs. 5.6 and 5.11, we see the basic problem we have in making inferences about the past from current levels of diversity. Both of these simple experiments lead to the same end results— after 4000 generations the level of mtDNA diversity in equal to 5.2 in population A and 3.8 in population B. In the first scenario (Fig. 5.6), this difference in diversity was due to a difference in population age—population A was older and, hence, more diverse. In the second scenario (Fig. 5.11), the difference in diversity was due to a difference in population size—population A was larger and, hence, more diverse. When we use genetic data to make inferences about the past, we do not know *previous* levels of diversity, only those from living populations (generation 0 in both graphs). We would know that population A was more diverse in the present day,

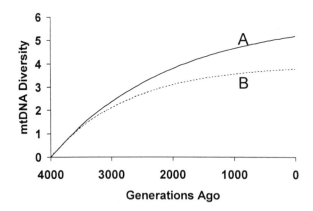

Figure 5.11. *Evolution of mitochondrial DNA diversity in two simulated populations, each for 4000 generations. The only difference between these two populations is that population A consists of 2000 reproductive females and population B consists of 1330 reproductive females. The difference in population size affects the increase in diversity each generation—larger populations will have higher levels of mtDNA diversity. As a result, population A shows more diversity when measured in the "present" (generation 0). See chapter notes for computational details.*

but we would not be able to tell whether this difference was due to a difference in population age or a difference in population size. We must deal with the relative likelihood of the underlying models.

I have mentioned that some researchers doubt that regional differences in diversity are a consequence of bottlenecks formed during phylogenetic branching. What of the alternative? How likely are regional differences in population size? More specifically, we need to examine the hypothesis that Africa was larger during much of recent human evolution, because Africa has the highest levels of genetic diversity. The proposition that Africa had the largest population may sound strange given the fact that Asia has the largest human population of all six inhabited continents. According to the mid-1999 estimates from the Population Reference Bureau,[23] the population size of Asia is over 3.6 billion. Africa is the second largest continent, with 771 million people, of which 630 million live in sub-Saharan Africa. The third largest continent today is Europe with 728 million. Given these regional differences, we might expect that Asia, rather than Africa, would have the greatest genetic diversity.

The problem with this conclusion is that these numbers refer to *present-day* population sizes, which are likely quite different from those earlier in history or prehistory. Population size and distribution have been affected by the rapid changes accompanying the origin and spread of agriculture over the past 12,000 years. In earlier times, when all humans existed as hunters and gatherers, much of the currently used land mass would not have been able to support large numbers. We cannot use present-day population sizes or distribution to make inferences about the past; our ancestors lived a different existence (hunting-gathering) up until only a short time ago.

Contrary to many people's expectation, Africa is a *large* continent. The area of Africa is almost 12 million square miles, making it the second largest continent in the world. We tend to lose sight of this fact because much of our awareness of differences in continental landmasses comes from looking at the standard Mercator projection, designed to represent *shape* accurately at the expense of accuracy in *size*. The standard Mercator projection does preserve the shapes of landmasses but introduces considerable error in portraying relative size. Greenland, for example, looks much larger than it really is. A glance at a Mercator projection map suggests that the former Soviet Union is larger than Africa when, in reality, it is smaller. A different type of map, using what is known as the Peter's projection, correctly shows the large size of the African continent.

Although large, the African continent is not the largest. Asia is larger, occupying roughly 17 million square miles. It might seem reasonable from these numbers to assume that Asia could fit more people and hence would have the largest human population in the past as well as the present. We need to keep in mind that these figures refer to *total* land mass and not to how much land mass is actually usable for hunters and gatherers. For one thing, not all of Europe and Asia would have been habitable during parts of human evolution because of recurrent ice ages. One estimate of land use for the Upper Paleolithic suggests that as much as 90% of Africa was occupied by humans but only 40% of Eurasia (this figure is estimated to have been even lower in earlier times).[24]

In addition, it seems reasonable to assume that population density was higher throughout sub-Saharan Africa than elsewhere in the Old World. Combined with the large land area, this might mean a larger human population in Africa than elsewhere. Archaeologist Fekri Hassan has estimated hunting-gathering density for different environments.[25] Although all hunting-gathering societies have low population densities, there is considerable variation across different environmental zones, ranging from a low of 0.01 persons per square kilometer in the Arctic to a high of 0.43 persons per square kilometer in subtropical environments (characteristic of sub-Saharan Africa). Hassan expresses these differences in an interesting manner that gives us a better idea of environmental variation in population size. To start with, Hassan suggests that most of the resources exploited by a hunting and gathering band will occur within a 10-km range of the base camp, a typical maximum value for foraging within a single day. A circle with a radius of 10 km would encompass an area of 314 km² (the area of a circle is equal to πr^2). Hassan then multiplied this area times the average population density for four environmental zones to obtain the number of people that could be supported by a 314-km² area: arctic = 3 people, semidesert = 11 people, grassland = 54 people, and subtropical savanna (the environment characteristic of much of Africa) = 136 people. The large landmass of Africa, combined with a higher percentage of usable land and higher densities for hunting-gathering populations, suggests that throughout much of human prehistory more humans lived in Africa than elsewhere in the Old World.[26]

The ecological and demographic considerations support the view that Africa would have had a larger population than other regions in the Old World. It is only very recently, with the advent of agriculture, that this situation was likely to change dramatically. Given a larger African population, and remembering that genetic diversity is proportional to population size, we would expect Africa to show the greatest amount of within-group diversity. To make this generalization, we must also consider the potential impact of gene flow, which can also increase genetic diversity. Could we also explain higher African diversity by greater gene flow *into* Africa than *out of* Africa? All other things being equal, populations that receive more gene flow will have greater genetic diversity. Is Africa more genetically diverse because of larger population size, greater rates of incoming gene flow, or both?

My colleague Henry Harpending and I addressed this question by modifying a method originally developed by Henry and Richard Ward for comparing observed and expected heterozygosities in a set of populations.[27] Observed heterozygosity is what I described earlier in this chapter. The expected heterozygosity of a population can be derived from the population's genetic distance from the average allele frequencies of a set of populations. Details of this method are given in the chapter notes. The key point is that observed and expected heterozygosities should be the same *under certain assumptions*. If, in any real world analysis, the observed and expected values are *not* the same, then the logical conclusion is that one or more of the underlying assumptions is not correct. This may sound confusing, but is actually a common method in science—we often learn a lot from cases where a model does *not* fit the observed data.

Harpending and I applied a variant of this method to a sample of craniometric data representing the major geographic regions of the Old World (described earlier and referenced in Fig. 5.4).[28] An advantage of the model underlying this method is that differences in gene flow between geographic regions are taken into account and will not affect the relationship between observed and expected variation. We found that the sub-Saharan African sample showed higher variation than expected from the model. The fact that the observed and expected variances are different can only be explained by one or more of the assumptions of the model being violated. When applied to worldwide data, our method makes three assumptions: 1. Mutation rates are the same in each region; 2. There is no gene flow from outside the region of study; and 3. Population sizes are the same in each region. We can easily justify the first two assumptions. For differences in mutation rates to be responsible, we would have to assume mutation rates that varied by several orders of magnitude across regions, something that is not possible. The second assumption, no external gene flow, is also easy to justify when considering the fact that our region of analysis was the world. Contrary to the plots of many science fiction stories, our planet has not received any extraterrestrial gene flow!

This leaves only the third assumption—equal population sizes. Because the expected variances are based on a model assuming equal population sizes and the observed variances are not the same as the expected variances, then clearly this assumption is violated. Using this standard form of hypothesis testing, we were able to show that the craniometric evidence supports regional variation in population size. In addition, the fact that the sub-Saharan African sample showed more variation than expected suggests that the population size of Africa was larger than for other geographic regions.

We could argue, of course, that any genetic inferences from craniometric traits might be biased because such phenotypic traits are affected partially by developmental and environmental influences, and some of these traits may not be neutral. Several years later, I engaged in a similar study with my colleague Lynn Jorde, but this time we used a dataset consisting of 60 microsatellite DNA loci.[29] These loci appear to be selectively neutral and are completely heritable. We found similar results—Africa showed more diversity than expected. The results of both craniometric and microsatellite DNA studies are summarized in Fig. 5.12, which presents the observed and expected levels of regional diversity based on the assumption that regional population sizes were equal. In both cases, Africa is more diverse than expected and the other geographic regions are less diverse than expected.

These studies show that the assumption of equal population sizes is rejected and it is likely that Africa had a larger population size than other regions. How much larger? Henry and I developed a simple, although computationally intensive, method to answer this question. We reran our analyses using different combinations of relative population size, which is the proportion of *total* population size across all regions. In our initial analysis, we had assumed equal population sizes for our four geographic regions, which means the relative population size for each region was 1/4 = 0.25. We examined a large number of different proportions for each region, such as

Europe = 0.4, Africa = 0.2, Asia = 0.3, Australasia = 0.1

Europe = 0.2, Africa = 0.2, Asia = 0.2, Australasia = 0.4

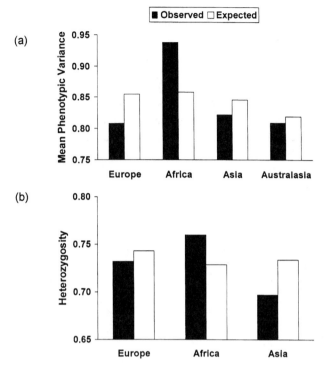

Figure 5.12. Observed regional variation compared to the variation expected under the hypothesis that the regions had equal population sizes. a. Observed and expected craniometric variation based on the average standardized variance for 57 craniometric variables (Relethford and Harpending 1994) b. Observed and expected heterozygosity based on 60 microsatellite loci (Relethford and Jorde 1999).

among many others. For each different combination, we examined the fit between observed and expected variances, looking for the combination of proportions that produced the best fit. Our best fit was found when we assigned Africa a relative weight of 0.5 and each of the other three geographic regions a relative weight of 0.167. In other words, observed and expected variances matched up when we assumed that Africa accounted for 50% of the total species.

Lynn Jorde and I applied the same method to his microsatellite DNA data from three regions in the Old World. The best fit occurred when we assigned relative weights of Africa = 0.73, Europe = 0.18, and Asia = 0.09. These results suggest that Africa accounted for almost three-fourths of the total species. Figure 5.13 shows the results of the craniometric and microsatellite DNA studies when the best fitting proportions were used—this graph clearly shows that observed and expected diversities are almost identical. Lynn and I also showed that the ranges of proportions providing acceptable fits to the data overlapped for the craniometric and microsatellite DNA analyses. A relative African size of 60% provides a reasonably good fit to both data sets. Regardless of the specific numbers, these studies show that higher African diversity can easily be explained in terms of regional variation in population size, a result that can be incorporated into both African replacement and multiregional models of modern human origins.

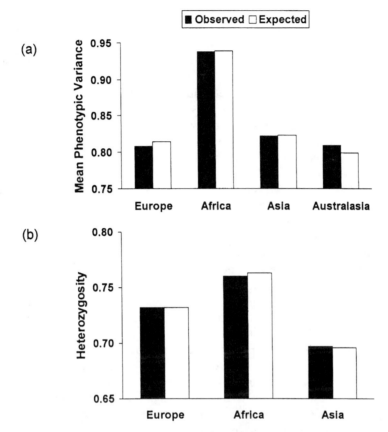

Figure 5.13. *Observed regional variation compared to the variation expected given the best fit of relative regional population sizes. a. Observed and expected craniometric variation based on the average standardized variance for 57 craniometric variables. The best fitting relative weights were Africa = 0.500, Asia = 0.167, Europe = 0.167, Australasia = 0.167 (Relethford and Harpending 1994) b. Observed and expected heterozygosity based on 60 microsatellite loci. The best fitting relative weights were Africa = 0.73, Asia = 0.09, and Europe = 0.18 (Relethford and Jorde 1999).*

The genetic evidence supports the hypothesis that throughout much of human prehistory more of our ancestors lived in Africa than elsewhere. This finding is based on population genetics models that make a number of simplifying assumptions that must be looked at in closer detail to relate these findings to the real world. For one thing, we are assuming that the ancestral African population actually lived in Africa, although it is possible that parts of the ancestral African population lived elsewhere. This suggestion has been made as one explanation for the discovery of both anatomically modern-humans and Neandertals in the Middle East.[30] Assuming for the moment that these two groups represent different populations (an assumption not everyone agrees with), there may have been alternating occupation of the Middle East. According to this idea, at one point there were moderns from Africa living in the Middle East while the Neandertals were living in Europe. During colder times, the moderns moved south back into Africa and the Neandertals moved south into the Middle East. At other times, the situation reversed, making the Middle

East the prehistoric equivalent of a time-share condominium! I am not particularly arguing for this model but offer it simply to illustrate that population movement, particularly those back and forth across continental lines, can complicate any simple interpretation we have regarding the fixity in time and space of biological populations.

Still, ecological inference combined with genetic data does make a strong case for a larger long-term African population size. The important qualifier here is "long-term" because genetic models frequently make use of a single estimate of population size, whereas in reality population size changes over time. The suggestion of a larger African population size does not mean that Africa was *always* larger than other regions (although it could have been) but rather that the *average* population size over time was larger. Archaeologist Richard Klein argues against the interpretation of larger African size because of evidence suggesting that Africa was actually lightly populated during the time of the emergence of modern humans.[31] It may be the case that there were specific times during which Africa did not have a larger human population than other regions, but this does not necessarily rule against models where the *average* long-term population was larger.

The key question at this point is how a larger African population fits different models of modern human origins. Both African replacement and multiregional models predict that our ancestors initially came from Africa; the difference is in the timing of this event. According to the African replacement model, modern humans arose in Africa as a separate species that later dispersed out of Africa to replace extant hominids throughout the rest of the Old World. Multiregional models also argue that our ancestors arose in Africa, but much earlier with the origin of *Homo erectus*, and gene flow continued to link humans in the Old World in a single evolving lineage. Both models have humans (in the broad sense of the word) starting in Africa and winding up spread out over the entire Old World. The difference between the models is in timing and the genetic relationship of regional populations over time. Both models must deal with the ecological restrictions on population size for hunting and gathering populations. Although cultural changes can (and probably did) change the maximum population density, it seems reasonable to assume that Africa could have supported larger numbers throughout much of prehistory. This situation would have remained the same until very recently, when the origin and spread of agriculture changed much of the basic adaptive nature of human beings.

The African replacement model is consistent with the suggestion of a larger long-term African population size. Under this model, our ancestors began in Africa as a new species, from which new founding populations split off to replace extant hominids outside of Africa. These founding populations would have been small to begin with, with later population growth. This later growth means that the *average* population size over time would have been smaller outside of Africa. This finding, combined with the ecological restrictions discussed above, shows that the African replacement model is perfectly consistent with larger African population size. According to this view, the finding of higher genetic diversity in Africa *does* support a recent African origin of our species, although for reasons relating to population size rather than the relative age of populations.

This does not *prove* the African replacement model, because larger African population size is also compatible with multiregional models for much the same reason. Milford Wolpoff and Rachel Caspari, two leading proponents of multiregional models, argue that "... as we believe, and the archaeological record seems to show, there were more people living in African for most of human prehistory, and human populations outside of Africa were smaller and fluctuated more because of the changing ice-age environments."[32] They further argue that multiregional models are not only compatible with a larger African population—they require it.

5.5 SUMMARY

A growing body of genetic data shows us that, on average, populations of living humans in sub-Saharan Africa are the most genetically diverse. Because contemporary genetic diversity is a reflection of past evolutionary events, this finding suggests that some factor(s) in the past led to regional differences in genetic diversity. The question is how to relate this finding to various models of modern human origins.

Greater African diversity in mitochondrial DNA and other traits has been argued as evidence for a recent African origin and replacement of non-African archaic populations. According to this view, genetic diversity is an index of population age. Because African populations show greater levels of genetic diversity, they must therefore be older and non-African populations must be younger. Because the African replacement model predicts the rise of modern humans first in Africa, with other geographic regions being occupied later in time, it therefore predicts that the greatest genetic diversity will be in Africa.

Although this makes sense at one level, the basic premise depends on a number of factors such as the assumption of severe and long-lasting bottlenecks and continued postdispersal isolation of regions. These assumptions need further examination to determine whether an appropriate choice of parameters is compatible with observed genetic variation. Even so, the problem will not be solved because another, more likely, explanation exists for regional differences in genetic diversity—regional differences in population size. Several studies have shown that greater African genetic diversity is easily explained by a larger average long-term African population size, a suggestion that fits what we know about the demography and ecology of early humans.

Both models are compatible with the view that humans, regardless of species, were more numerous in Africa than elsewhere. The difference between the models is *which* of these humans were our ancestors. The good news is that both models can relate higher African genetic diversity with a larger long-term African population size. This is good news because we can feel confident of this demographic and ecological finding regardless of which specific origin model is correct. The bad news, however, is the same as the good news. Because both models are compatible with the same observation, we have no way to choose between them based solely on the genetic evidence of regional levels of genetic diversity. As with the case of estimating dates for our most recent common ancestor (Chapter 4), there are alternative ways of interpreting the evidence of regional diversity. In both cases, we have learned a lot about the past demographic history of our species, although the answer to the specific question of *who* our ancestors were eludes us. Again, it is necessary to consider further evidence before trying to pull all of the pieces together. This chapter has focused on genetic differences *within* living human populations. The next chapter proceeds to the next type of genetic evidence pertaining to modern human origins—the pattern of genetic differences *between* living human populations.

Chapter 6

Genetic Differences Between Human Populations

Much of evolutionary theory deals with the meaning of differences between groups. Chapter 5 considered variation *within* living human populations; this chapter will examine the evidence for genetic differences *between* living human populations and what these differences tell us about our past. There are two basic issues concerning human genetic differences that have implications for the modern human origins debate—the average *level* of genetic differences between populations and the *pattern* of genetic differences between populations. Both of these concerns are directly related to the concept of genetic distance.

6.1 GENETIC DISTANCE AND EVOLUTION

We are all familiar with the concept of distance in a geographic sense and can use measures of geographic distance to solve basic problems. For example, which is closer to Athens, Greece, London, England, or Tel Aviv, Israel? To answer this, we need to know the geographic distance between these cities. The straight-line distance between Athens and London is 2391 km, and the straight-line distance between Athens and Tel Aviv is 1215 km. Clearly, Athens is closer to Tel Aviv than to London because the distance is smaller. The smaller the geographic distance between two cities, the closer they are in space. As illustrated by the examples in Chapters 1 and 2, the same general principle holds with genetic distance—the smaller the genetic distance between two populations, the closer they are genetically.

A genetic distance is simply a measure of how different two populations are genetically. For example, imagine that you have studied three populations (A, B, and C) for a particular allele and obtained the following frequencies:

$$\text{population A} = 0.3$$

$$\text{population B} = 0.8$$

$$\text{population C} = 0.7$$

What could you say about the genetic relationship of these three populations? In particular, can you tell which two of these populations are the most similar for this allele? It is obvious that populations B and C are more similar to each other than either is to population A because their allele frequencies lie closer to each other. One way to see this is to examine the absolute difference in allele frequencies between each pair of populations—subtract the smaller frequency from the larger. When comparing population A to population B, the absolute difference in allele frequencies is $0.8 - 0.3 = 0.5$. The absolute difference between populations A and C is $0.7 - 0.3 = 0.4$. The last comparison, between populations B and C, gives an absolute difference of $0.8 - 0.7 = 0.1$. We now list these differences for easy comparison:

populations A and B:	0.5
populations A and C:	0.4
populations B and C:	0.1

It is clear that populations B and C are the most similar to each other because the difference in allele frequencies (0.1) is the lowest. Population A is more distant from both populations B and C.

There are many different genetic distance measures, most of which are highly correlated with each other. In actual studies, we do not use the absolute difference in allele frequencies, but the basic principle remains the same—pairs of populations with smaller genetic distances are more closely related to each other. Many commonly used genetic distance measures look at the *squared* difference between allele frequencies, which is related to the covariance of allele frequencies around a set of mean allele frequencies in a given study. One example of a squared genetic distance is the Harpending-Jenkins distance. When comparing population i with population j, the Harpending-Jenkins distance is defined as

$$\frac{(p_i - p_j)^2}{\bar{p}(1 - \bar{p})}$$

where p_i and p_j refer to the allele frequencies of populations i and j. The denominator of this equation, $\bar{p}(1 - \bar{p})$, is a method used for standardization. The symbol $\bar{p}$ is the average frequency—as noted in Chapter 5, this is usually computed as a weighted average where each population is weighted by its population size (in this example, we assume equal population sizes).[1]

As an example, consider the allele frequencies for the three hypothetical populations we have been discussing: 0.3, 0.8, and 0.7. To use the Harpending-Jenkins distance method, we must first compute the average allele frequency, which is $(0.3 + 0.8 + 0.7)/3 = 0.6$. Once we have this number, we can compute the genetic distances between all pairs of populations. The genetic distance between populations A and B is

$$\frac{(0.3-0.8)^2}{0.6(1-0.6)} = \frac{(-0.5)^2}{0.6(0.4)} = \frac{0.25}{0.24} = 1.04$$

Likewise, the genetic distance between populations A and C is

$$\frac{(0.3-0.7)^2}{0.6(1-0.6)} = \frac{(-0.4)^2}{0.6(0.4)} = \frac{0.16}{0.24} = 0.67$$

and the genetic distance between populations B and C is

$$\frac{(0.8-0.7)^2}{0.6(1-0.6)} = \frac{(0.1)^2}{0.6(0.4)} = \frac{0.01}{0.24} = 0.04$$

Once again, we see that the smallest genetic distance is between populations B and C (0.042), whereas population A is more distant from both B (1.04) and C (0.67). The smaller the genetic distance between two populations, the more similar they are genetically. Of course, we would not want to base any inferences on only one allele; in practice, we compute the genetic distances for dozens and dozens of alleles and then average them.

The analysis of genetic distances can tell us which populations are more similar or more dissimilar. In the example above, population A was the most different. Why? There are a number of reasons a population might be more genetically different. It could be a reflection of historical relations between populations. In this case, perhaps populations B and C shared a more recent common ancestry. Imagine, for example, that population A splits at some point in the past, and part gave rise to another population, which later split into populations B and C. In this scenario, populations B and C have a more recent common ancestor, and we would therefore expect them to be more genetically similar to each other than to the more distant population A (see Fig. 6.1).

If the fissioning of populations, referred to here as phylogenetic branching, were the only way to generate genetic differences between populations, then we could take any table of genetic distance measures and quickly reconstruct the history of a species. The phylogenetic branching model works well when considering genetic distances *between species* (as described in Chapter 2) but is not the only (or even the most appropriate) model to use when considering genetic distances between populations *within* a species.

Another approach, referred to here as the population structure approach, considers genetic distances between populations to be a function of gene flow and genetic drift. In terms of the hypothetical example, populations B and C might be more similar genetically because they have shared more genes through migration. Population A could be more genetically distant because of less gene flow with either B or C or because it experienced more genetic drift, among other possible reasons. A key feature of this approach (see Fig. 6.2) is that there are not differences in the "age" of populations. All populations have the same time depth.

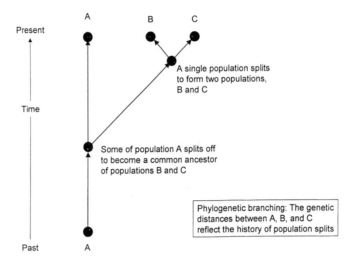

Figure 6.1. *Diagram of the phylogenetic branching model in which regional populations form through the process of daughter populations splitting off from a parental population in such a manner that genetic distances between populations in the present reflect the past history of these splits.*

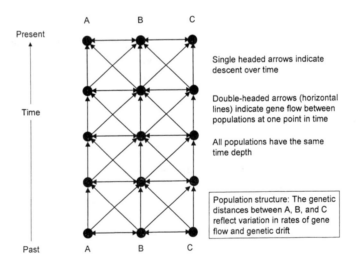

Figure 6.2. *Diagram of the population structure model in which all regional populations share similar time depth and are not isolated but connected to each other via gene flow. Genetic distances between populations in the present reflect the history of gene flow and genetic drift.*

6.2   LEVELS OF GENETIC DIFFERENTIATION

One useful genetic distance measure focuses on the overall level of genetic differentiation, which is a summary measure of differences across all populations. A commonly used measure of overall differentiation is Wright's F_{ST}.[2] Instead of focusing on genetic distances between pairs of populations, F_{ST} is a measure of the average genetic distance to the mean of all populations (often referred to as the "centroid").

There are several different ways of estimating F_{ST}. One of these methods expresses F_{ST} in terms of genetic distance, making it a useful method for illustrating exactly what differentiation means.[3] We start by looking at the genetic distance of each population to the mean, using the same formula shown earlier, but using the mean allele frequency ($\bar{p}$) in place of one group:

$$\frac{(p_i - \bar{p})^2}{\bar{p}(1 - \bar{p})}$$

As an example, consider the allele frequencies for hypothetical populations A, B, and C presented earlier (A = 0.3, B = 0.8, C = 0.7) with a mean of 0.6. The genetic distance of population A to the mean is

$$\frac{(0.3 - 0.6)^2}{0.6(1 - 0.6)} = \frac{(-0.3)^2}{0.6(0.4)} = \frac{0.09}{0.24} = 0.375$$

Similarly, the genetic distance of population B to the mean is

$$\frac{(0.8 - 0.6)^2}{0.6(1 - 0.6)} = \frac{(0.2)^2}{0.6(0.4)} = \frac{0.04}{0.24} = 0.167$$

and the genetic distance of population C to the mean is

$$\frac{(0.7 - 0.6)^2}{0.6(1 - 0.6)} = \frac{(0.1)^2}{0.6(0.4)} = \frac{0.01}{0.24} = 0.042$$

F_{ST} is simply the average genetic distance to the mean, which in this case is

$$\frac{0.375 + 0.167 + 0.042}{3} = \frac{0.584}{3} = 0.195$$

Normally, we would take an average weighted by population size, but in this example, I assumed all three populations had the same size just to make things easier to compute.

The higher the value of F_{ST}, the more distant (on average) populations are from the mean allele frequencies. Thus F_{ST} provides a useful index of how spread out populations are genetically. Figure 6.3 provides a simple visual analogy of high and low levels of F_{ST}. Low values of F_{ST} correspond to cases in which populations are not very different genetically from one another or from the mean. High values of F_{ST} indicate situations in which populations are more genetically divergent from one another and the mean.

The value of F_{ST} reflects a balance between mutation, gene flow, and genetic drift. Mutation and gene flow act to reduce genetic differences between populations, and genetic drift acts to increase genetic differences. As shown in Fig. 6.4, F_{ST} will

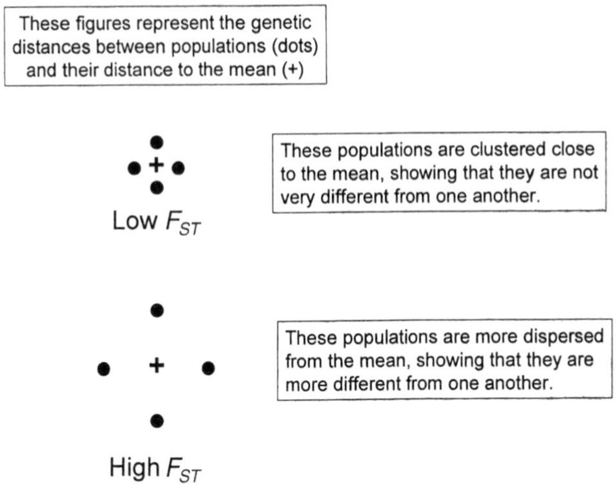

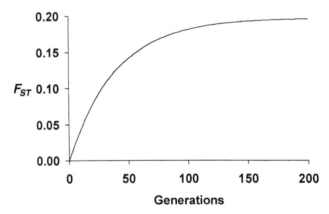

These figures represent the genetic distances between populations (dots) and their distance to the mean (+)

These populations are clustered close to the mean, showing that they are not very different from one another.

Low F_{ST}

These populations are more dispersed from the mean, showing that they are more different from one another.

High F_{ST}

Figure 6.3. *Graphic representation of F_{ST}, the relative amount of among-group differentiation. Low values of F_{ST} reflect sets of populations that have not genetically diverged. High values of F_{ST} reflect greater genetic divergence. The level of F_{ST} depends on the time depth of the populations, population size, and rates of gene flow and mutation.*

Figure 6.4. *Simulation of the evolution of F_{ST} over time among a set of populations of 100 reproductive adults, a rate of gene flow of 0.01 per generation, and a mutation rate of 0.00001. See chapter notes for computational details.*

increase over time to reach an equilibrium where the increase in genetic distance (due to drift) is at a balance with the decrease in genetic distance (due to mutation and gene flow).[4] Comparison of F_{ST} values can tell us about differences in population age, rates of gene flow, and population size. All other things being equal, and assuming the populations have not reached equilibrium, the younger a set of populations, the lower the value of F_{ST}. If two sets of populations are of the same age, then higher rates of gene flow and/or larger population sizes will be reflected in lower levels of F_{ST}. Likewise, lower rates of gene flow and/or smaller population sizes will be reflected in higher levels of F_{ST}.

A number of studies have estimated F_{ST} based on regional populations across the world. Some of these studies are difficult to compare because different numbers and definitions of geographic regions have been investigated, and F_{ST} is dependent to some extent on the number of groups included in an analysis. Some researchers have focused on three regional groups (Africans, Europeans, Asians), whereas others have included other regions such as Australasia and the New World. Despite this variation in methodology, the results from studies analyzing between three and six geographic regions are very consistent—F_{ST} in the human species is roughly 0.10 to 0.12. This value has been found for classical genetic markers, nuclear DNA RFLPs, some microsatellite DNA loci, *Alu* insertions, and craniofacial measurements.[5] F_{ST} values for some microsatellite loci and mtDNA often tend to be lower, perhaps reflecting higher mutation rates for these loci.

An F_{ST} of 0.10 is very low compared to other mammalian species. Alan Templeton looked at values of F_{ST} from studies of buffalo, waterbuck, impalas, wildebeests, bighorn sheep, coyotes, elephants, gazelles, and wolves.[6] He found that only African buffalo, waterbuck, and impala populations have similar, or lower, F_{ST} values than humans. The remainder all had higher F_{ST} values. Wildebeest, bighorn sheep, and coyote populations had F_{ST} values between 0.2 and 0.3, whereas the remainder of the species analyzed had values greater than 0.4.

The low F_{ST} value for the human species shows an interesting feature of human biological diversity—we are a rather homogeneous species. Although genetic differences exist between human populations across the world, the relative magnitude of the variation among populations is low, particularly when compared to other species. Another way to look at F_{ST} is as the proportion of total genetic variation due to variation across groups. When we divide the human species into different geographic regions (e.g., Africa, Europe, etc.), we are often interested in how this division relates to the total genetic variation in the species. An F_{ST} of 0.10 means that only 10% of the species' genetic diversity is due to differences *between* geographic regions. The remaining 90% of the diversity exists *within* geographic regions. More elaborate analyses have found that roughly 10% of the total genetic variation in the human species occurs *between* geographic regions (what some call "races"), 5% occurs between populations within regions, and the remainder (85%) occurs within local populations.

The notion that humans can be subdivided into races permeates much of the past intellectual history of Western civilization. The biological race concept suggests that a species can be easily subdivided into discrete groupings, each of which is relatively homogeneous. If so, more biological variation should exist *between* groups and much less *within* groups. We see the remnants of this type of thinking when people suggest that "they look different from us" (variation *between* groups) and "they all look alike" (variation *within* groups). Although this is the case for some species, the concept breaks down for humans. If we fit the biological race model, we would expect to see more of the variation existing between geographic regions and much less within these regions. Instead, we see exactly the opposite. Part of our confusion about human diversity comes from the fact that most schemes of racial classification rely heavily on skin color, a trait that *does* show a lot of variation between regions.[7] However, the F_{ST} for skin color is atypical, having been shaped through natural selection to varying levels of ultraviolet radiation across our planet.

What does a low F_{ST} tell us about modern human origins? Unfortunately, there are several possible answers, each of which supports a different origin model. One possibility is that the low F_{ST} reflects insufficient time for regional differences to have developed (recall from Fig. 6.4 that F_{ST} increases over time). If so, then we might be seeing the genetic signature of a relatively recent origin of regional differences, which fits the African replacement model. On the other hand, a low F_{ST} could also reflect a relatively large number of migrants into each region. How much migration is needed to obtain human F_{ST} values?

F_{ST} is related to the total number of migrants into any population in a single generation (M) as

$$M = \frac{(1 - F_{ST})(g - 1)^2}{4g^2 F_{ST}}$$

where g is the number of populations.[8] As an example, consider an analysis performed by Gregory Livshits and Masatoshi Nei.[9] They computed a value of F_{ST} = 0.114 between three regional populations (Africans, Europeans, Asians) using data on 86 blood protein and enzyme loci and 33 blood group loci. Using the above equation with F_{ST} = 0.114 and g = 3, we obtain M = 0.86 migrants into each population every generation. This number refers to the *total* number into each population, regardless of where they came from. We can get a rough idea of the magnitude of migration between pairs of populations by dividing this number by $g - 1$, which gives 0.86/(3 − 1) = 0.43 migrants per generation.

Most studies show an F_{ST} between 0.10 and 0.12 when three to six geographic regions are used. This range in F_{ST} and g means that the total number of migrants (M) into any region in a given generation will range from roughly 0.8 to 1.6. The number of migrants exchanged between regions in each generation ranges roughly from 0.25 to 0.50, depending on the number of regions being considered. These numbers correspond to one migrant moving from one region to another every two to four generations. Of course, such estimates make a number of simplifying assumptions. For one thing, the model above ignores mutation. If mutation is added to the model, and we are considering a trait with a high mutation rate, then the number of migrants needed to produce a given F_{ST} value will be lower.

The simple model also assumes that regional population sizes are equal, which is clearly not the case. As discussed in Chapter 5, there is strong evidence that Africa had the largest long-term average population size. Methods that are more complex exist to estimate migration under such conditions. Henry Harpending and I looked at genetic distances in classical genetic markers and craniometrics between four geographic regions under a model of larger African population size, estimating 0.33 migrants exchanged between each region each generation.[10] This level of migration is equivalent to one migrant exchanged every three generations.

This estimate is an oversimplification because it assumes that migration was constant over time; that is, 0.33 migrants *each* generation. In reality, we would expect migration to change over time, reflecting both periods of relative isolation as well as periods of rapid movement, perhaps both correlated with periods of climactic change. The estimates discussed here are simply averages; the same result could be obtained by a larger number of migrants exchanged less frequently. One migrant every third generation is equivalent to 10 migrants every 30 generations or 20 migrants every 60 generations. It is perhaps more reasonable to think of small groups migrating over long distances infrequently than a single migrant more frequently. Estimates derived from genetic data serve to describe average patterns; archaeological, paleontological, and environmental data are needed to work out the specifics. In any case, these rough estimates show us that we do not require very large levels of migration to produce a relatively low F_{ST} value. The relative homogeneity of the human species might have been the product of a recent African origin, but it could also be due to gene flow over time under either origin model.

6.3 THE PATTERN OF GENETIC DISTANCES BETWEEN POPULATIONS

Estimating F_{ST} tells us about the *magnitude* of genetic differentiation. It provides information about the average genetic distance to the centroid. It does not tell us anything about the *pattern* of genetic distances. To determine pattern we need to know the genetic distances between each pair of populations. The simple example shown in Fig. 6.3 has equal distances between each pair of populations. This is not likely to be the case in any actual study of genetic distances, where some populations will be more similar to others.

Studies of genetic distances between human populations almost all show the same basic pattern—Africa is more genetically divergent. Other regional populations, such as Europe and Asia, are more similar genetically than either is to Africa.[11] Several examples illustrate this general finding. My colleague Henry Harpending and I computed genetic distances between four geographic regions (Africans, Europeans, East Asians, Australasians) based on 93 alleles at 37 classical genetic marker loci. The best way to interpret genetic distances is to draw a picture that represents graphically the overall pattern of genetic distances. There are a number of ways to do this; one commonly used method is cluster analysis, which produces a treelike picture with genetically similar populations being placed in the same cluster. Figure 6.5 provides the cluster diagram (known as a dendrogram) of our genetic distance matrix.[12] Populations that are joined together in clusters close to the bottom of the graph are more genetically similar. Less similar populations join a cluster towards the top of the dendrogram. East Asians and Australasians are the most genetically similar, followed by Europeans. Africans are the most genetically divergent.

Figure 6.6 shows a similar pattern. This dendrogram was based on genetic distances published by Anne Bowcock and colleagues. These distances were computed from allele frequencies of 99 RFLP loci for 5 populations. Africa (represented by pygmies from Central Africa and Zaire) is the most genetically distant.

Figure 6.7 shows the results of another genetic distance analysis by Cavalli-Sforza and colleagues, based on 120 classical genetic marker loci for 9 regional populations across the world. Moving up from the bottom of the dendrogram, we see populations clustering into easily identifiable geographic regions—Europe-Middle East-North Africa, Northeast Asia, Southeast Asia-Australasia-Oceania. Native Americans cluster with Northeastern Asians, in line with other evidence that Native Americans originated there. Once again, Africa is the most genetically divergent.

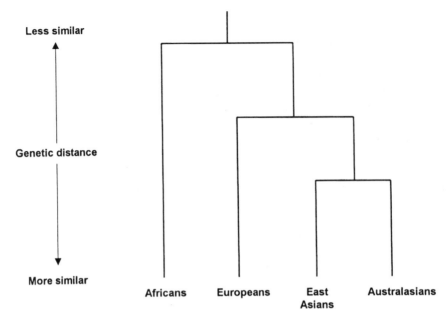

Figure 6.5. Dendrogram of genetic distances between four regional population clusters based on 93 alleles for 37 classical genetic marker loci (Relethford and Harpending 1995).

Many other studies show the same basic finding. There are differences between studies in the relative placement of non-African populations. Some show Europeans and Asians clustering together, followed by Australasians and/or Pacific Islanders (see, e.g., Figs. 6.6 and 6.7), whereas others show greater similarity between Asians and Australasians (see, e.g., Fig. 6.5). The exact dendrogram is a function of the number and nature of the population units sampled, the number of genetic traits, the method used to compute genetic distances, and the method of clustering. Despite such differences, virtually every study shows Africa to be genetically the most different. Keep in mind that we are looking at relative differences here. The level of genetic variation among groups (F_{ST}) for humans is relatively low. The *pattern* of genetic distances shows that, within this range of variation, geographic regions are not equidistant from one another. Africa stands out. The question is why.

6.4 EVOLUTIONARY INTERPRETATIONS

Why is sub-Saharan Africa genetically more distant than other geographic regions? The greater genetic divergence of Africa is not new—it was noted in studies over 25 years ago. In one early analysis of genetic distance, Yoko Imaizumi and colleagues noted some distinctiveness of African populations and suggested that this separation reflected the isolating effect of the Sahara Desert on gene flow.[13] Whether this is the case or not, this suggestion is firmly embedded in a population structure approach in which the pattern of genetic distances is seen as a reflection of gene flow and genetic drift. With the development of the African replacement model in the late 1980s, the focus of worldwide studies shifted toward the phylogenetic

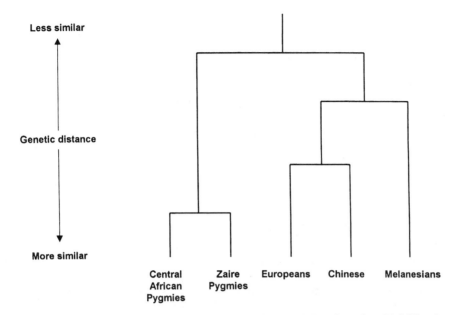

Figure 6.6. Dendrogram of genetic distances between five populations based on 99 DNA polymorphisms (Bowcock et al. 1991).

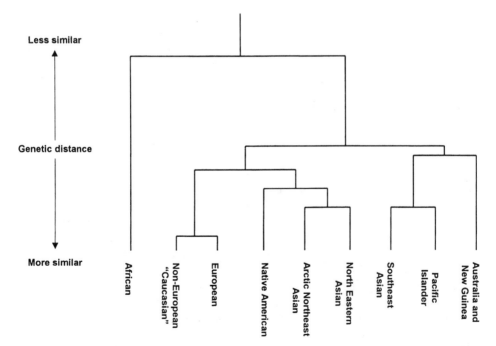

Figure 6.7. Dendrogram of genetic distances between nine population clusters based on 120 alleles for classical genetic markers (Cavalli-Sforza et al. 1994).

approach, in which the pattern of genetic distances is seen as a reflection of past branching of populations.

6.4.1   Do the Genetic Distances Reflect Phylogenetic Branching?

If we assume that a phylogenetic branching model (Fig. 6.1) is correct, then we would expect to get the same basic pattern of genetic distances we see in living populations—Africa would be the most distinct region. This expectation fits the African replacement model. After the initial emergence of modern humans in Africa 200,000 or so years ago, the first branching event would be the dispersal of some populations from Africa roughly 100,000 years ago, most likely moving through the Middle East before subsequently branching off to give rise to modern human populations in Europe, Asia, and elsewhere.

From a phylogenetic branching perspective, populations that have branched off more recently should be more genetically similar. Because many genetic studies show European and Asian populations to be more similar to each other than to the African population, the logical inference is that Europe and Asia split off from a common ancestor more recently than either did from Africa. This inference is analogous to the kinds of genetic relationships you would expect within your own family. You are more genetically similar to your sibling than to your third cousin because you and your sibling share a more recent common ancestor. The difference between this example and the inferences made in many studies of genetic distance is that you are looking at individual ancestors in your family, whereas genetic distance studies look at entire populations and their historical relatedness.

The phylogenetic branching approach is commonly used in many areas of evolutionary biology. The discussion in Chapter 2 regarding the common ancestor of humans and chimpanzees is an example of this same approach. The underlying assumption, that genetic distance reflects common ancestry, is appropriate for comparing species, whose historical relationship resembles a branching "tree"—species arise from other species and once separate remain separate. Assuming we are looking at neutral traits, we need only consider mutation and genetic drift as the causes of evolutionary changes within species over time that increase the genetic distance between species. Under this model, genetic distance is proportional to the time elapsed since two species diverged from a common ancestor.

There are many sophisticated models for analyzing genetic distances between species to reconstruct phylogenetic history. The basic problem when working with data from living humans is that we are a single species. Even if there were multiple human species in our past, the fact remains that there is only *one* human species today. It takes more than one species to reconstruct phylogenetic branching. When we compare, for example, human genetic data with that from chimpanzees and baboons, we evaluate possible phylogenetic trees to link these three species. If we have only one species, then there is nothing to link it to. How can we use the phylogenetic branching model to reconstruct the evolution of the human species when we have only one human species today?

Because living humans do not constitute separate species, some researchers have taken geographic regions as units of analysis, which leads to several problems. The choice of geographic boundaries is by nature somewhat arbitrary (where, for example, would you draw the line between European and Asian regions in the single land mass of Eurasia?). Perhaps the biggest problem is that a phylogenetic branching model assumes that these regions represent separate evolutionary entities. This is clearly not the case, because human populations are and have been interconnected by gene flow. Even if we did arise as a new species in Africa 200,000 years ago, and even if regional populations were at times isolated from each other, we have ample evidence that there has been gene flow for much of recent human evolution. In this case, a simple phylogenetic branching model is not valid.

Given these problems, what kind of inferences can we draw from Figs. 6.5–6.7? Given continued gene flow, we would not want to assign actual dates to the branches. Gene flow would act to make the branches appear more recent than they actually were, even assuming a branching history is the appropriate model.[14] Still, one could argue that such pictures represent the overall pattern of phylogenetic branching, regardless of our inability to date the branches as we did when comparing humans and chimpanzees. After all, these pictures certainly *look* like trees. This brings us to a basic and often overlooked problem—the fact that genetic distances produce a pattern that looks like a tree does not mean that a tree model is correct. The appearance of a tree is a consequence of the basic method of cluster analysis—it is *designed* to reproduce a tree of relationships. The "tree" in this case refers simply to the physical appearance of genetic distance relationships, and it is not necessarily a phylogenetic tree.[15]

As an example of this problem, consider Fig. 6.8, which shows a dendrogram of the genetic distances between five populations labeled by code letters (B, E, C, G, L). The overall pattern of distances is quite clear—there are two major clusters, one consisting of populations B and E and the other consisting of populations C, G, and L. *If* we decide to interpret this tree as a phylogenetic tree, then our logical

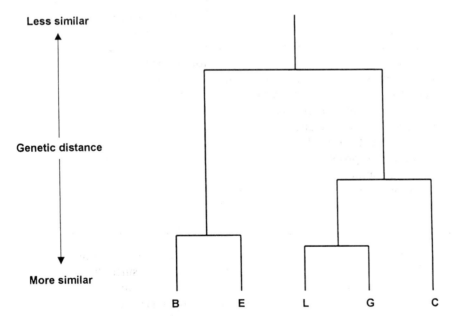

Figure 6.8. *Dendrogram of genetic distances between five populations. See text for the identity of these populations.*

conclusion would be that there was first a split into two clusters (B-E and C-G-L), each of which split further later in time. This would be a reasonable conclusion *if* we knew that these populations had indeed evolved under a phylogenetic branching model. Is this an appropriate assumption? There is no way for you to answer this question without knowing more about these five populations than their code letters. If we knew that these five populations represented different species, then a phylogenetic branching history would automatically be the most appropriate, and our cluster tree would provide a means of reconstructing that history. If these five groups represented isolated populations within the same species, a phylogenetic branching model *might* be appropriate, although other models involving past history of migrations and fissions might be more accurate. In this case, neither of these models is correct.

These five groups represent five small human populations along the west coast of Ireland. All five populations were studied in the late 1890s, at which time anthropometric measurements were collected. I derived the phenetic distances using new methods of making population genetic inferences from metric traits.[16] Three of the populations (C, G, L) were located within 13 km of one another along the west coast of County Galway in Ireland. The other two populations (B, E) were located near one another in County Mayo to the north of County Galway. The average distance between the populations in the two counties is 91 km. With this information in mind, Fig. 6.8 becomes immediately understandable; the division into two clusters reflects geographic separation, with each cluster corresponding to a county. This geographic correlation extends even further in the analysis. Note that populations L and G are more similar to one another than either is to population C. This pattern is also a reflection of geography, because populations L and G are geographically closer (4 km) than either is to population C (an average of 10 km). A strong relationship of genetic and geographic distance implies that the isolation by distance model, where gene flow (and hence genetic similarity) declines with geographic distance, is the most appropriate explanation for this specific data set.

My main point here is that inferences cannot be made exclusively from any graphical representation of genetic distances. More information is needed to place these patterns into an appropriate evolutionary context. We need to consider possible alternatives when dealing with the pattern of larger African genetic distances. Once again, the data are compatible with an African replacement model, which predicts initial origins in Africa and secondary dispersals out of Africa. The model predicts the type of pattern we have seen in Figs. 6.5–6.7. Once again, we cannot use this compatibility as proof of the model unless we can show that multiregional alternatives are *not* compatible with the genetic data. Can the same pattern of genetic distances be obtained under a population structure model?

6.4.2 Do the Genetic Distances Reflect Differences in Gene Flow?

If the African replacement model is correct, and the genetic distances reflect a history of phylogenetic branching, it is clear that gene flow since the initial dispersal out of Africa could change the genetic distances and conceivably distort our historical inferences. The degree to which this could happen is debatable, but a number of researchers argue that the primary split between African and non-African populations supports a recent African origin. Even though gene flow would act to make populations appear more similar than expected based solely on branching followed by isolation, this change could have affected all distances proportionately. Although the absolute genetic distances would have been affected, the *relative pattern* could remain the same.

Once we acknowledge that gene flow must be rightfully taken into consideration, and that perpetual regional isolation is not the appropriate model, we open up another question—could gene flow and genetic drift alone produce the same pattern of genetic distances? Might the larger African genetic distances be a reflection of population structure that *mimics* an African replacement model? Is it possible to obtain the same genetic distances under a population structure model without invoking any phylogenetic branching?

A basic problem in interpreting history from genetic distances is that the same result can often be reached by a number of alternative models. Joseph Felsenstein expresses this problem well in his discussion of problems of inferring history from genetic distances. He notes

> Population geneticists faced with geographical data on gene frequencies usually resort to clustering algorithms applied to a matrix of genetic distances. These inevitably result in a tree (or "dendrogram"). This gives the appearance of verifying that the species is hierarchically subdivided. Yet it may be that the genetic distances reflect isolation by distance, or some more complex pattern of gene flow, rather than a sequence of historical branching events.[17]

In terms of genetic distances among regional populations, this means that the greater genetic distances to Africa might not reflect a history of branching, but rather less gene flow between Africa and other regions.

A hypothetical example helps illustrate the relationship between gene flow and genetic distance. The key parameter is the *number* of migrants that are exchanged between populations. Consider three populations (labeled A, B, and C) that each consist of 1000 reproductive adults. Assume that these populations remain the same size from one generation to the next. Let populations A and B exchange one migrant each generation; that is, one person moves from A to B while one moves from B to A. Let populations A and C also exchange one migrant each generation, and let populations B and C exchange five migrants each generation. Figure 6.9 illustrates this pattern of migration.

We now need to figure out the number of people in each population that are *not* migrants. Population A consists of 1000 people, of which one person is exchanged with population B and one person is exchanged with population C. This means that there are two migrants each generation into population A, leaving 1000 − 2 = 998 nonmigrants. We do the same thing with population B—out of 1000 people there are six migrants (1 from A and 5 from C), leaving 1000 − 6 = 994 nonmigrants. There is the same number of nonmigrants in population C. The pattern of migration is best shown using a migration matrix, a table that easily shows the pattern of migration. In terms of this example, the migration matrix is

	A	B	C
A	998	1	1
B	1	994	5
C	1	5	994

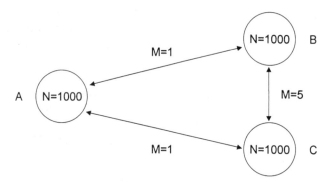

Figure 6.9. *Model of gene flow and genetic drift for three populations (A, B, C) used to derive the genetic distances shown in Fig. 6.10. Each population consists of 1000 reproductive adults. Populations A and B exchange one migrant each generation, populations A and C exchange one migrant each generation, and populations B and C exchange five migrants each generation.*

The rows of a migration matrix show where people came from in the previous generation, and the columns indicate where they wind up. For population A, we see that 998 people in A came from A, one person in A came from B, and one person in A came from C. For population B, one person came from A, 994 stayed in B, and five people came from C. For population C, one person came from A, five people came from B, and 994 stayed in C. Given this pattern, we would expect to see populations B and C to be more similar to each other genetically, because they exchange a larger number of migrants (5 each generation) than either exchanges with population A (1 migrant each generation).

This matrix refers to the *number* of migrants (and nonmigrants). We can convert these numbers into *rates* of gene flow by dividing each element in the matrix by the sum of each column. For example, looking at the column labeled A, we see that population A consists of 998 nonmigrants, one migrant from B, and one migrant from C, for a total of 1000 people. We would divide each number by this total (1000). We would then divide each number in column B by its total (which also is 1000 in this case), and finally divide each number in column C by its total (also equal to 1000). The resulting matrix is

	A	B	C
A	0.998	0.001	0.001
B	0.001	0.994	0.005
C	0.001	0.005	0.994

The elements of this matrix represent the probability that a gene in a given column came from a given row. For example, we see that the probability in column C for row B is equal to 0.005, which means there is a probability of 0.005 that any given gene in population C came from population B. These numbers are easily interpreted as *rates* of gene flow. We see that the highest level of gene flow between populations is between populations B and C (0.005), which is greater than the level of gene flow between populations A and B (0.001) or between populations A and C (0.001).

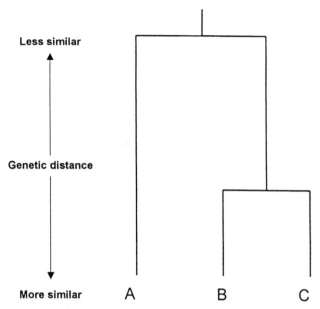

Figure 6.10. *Dendrogram of expected genetic distances between three hypothetical populations using the migration model in Fig. 6.9.*

Therefore, we would predict that, based on gene flow, populations B and C should be more similar genetically to each other than either is to population A.

To predict accurately the genetic distances between populations, we also need to factor in genetic drift.[18] I have done this using a method developed by Alan Rogers and Henry Harpending. Figure 6.10 shows the dendrogram of these predicted distances. It is clear from this picture that populations B and C are more genetically similar than either is to population A. The reason for this similarity is a higher amount of gene flow between populations B and C than with population A. Suppose we did not have the detailed history I used in this example but were instead trying to make inferences from Fig. 6.10. We might be tempted to explain this pattern of genetic distances by a phylogenetic branching model, in which populations B and C split from population A later in time. This example shows that the same pattern of genetic distances could also be explained by differences in gene flow without any population branching at all.

It is clear that although the pattern of regional genetic distances shown in Figs. 6.5–6.7 could be the result of a history of phylogenetic branching, they could also be due to variation in interregional gene flow. In particular, the genetic divergence of Africa may reflect lower rates of gene flow with other regions. Although it is tempting to interpret cluster trees as actual phylogenetic trees, we must always acknowledge the possibility that this is not the correct interpretation. The possibility of variation in interregional gene flow is perhaps better illustrated by using another method of graphic representation of genetic distances. Figure 6.11 offers a different way of portraying the genetic distances between nine geographic regions that were shown in Fig. 6.7. This graph is a principal coordinates plot, a method that

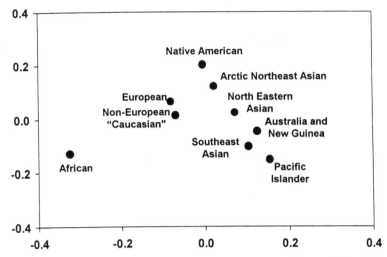

Figure 6.11. Principal coordinates plot of genetic distances between nine population clusters based on 120 alleles for classical genetic markers (Cavalli-Sforza et al. 1994). This plot is an alternative representation of the same genetic distances used in Fig. 6.7.

can reduce a matrix of genetic distances to a two-dimensional "map" in which genetically similar populations plot closer to one another.[19] Such maps are read the same way you would read a road map. Looking at a road map, you correctly interpret that the closer two points are on the map, the closer they are geographically. A genetic distance map provides a related inference—the closer two points are on the map, the closer they are genetically.

The genetic distance map in Fig. 6.11 shows the genetic divergence of Africa—it plots over to the side and away from other clusters. Northeastern Asian populations (and their descendant population, Native Americans) cluster together, as do Southeast Asian, Australasian, and Pacific Island populations. Note that this map does not convey any new information that we did not see in the dendrogram in Fig. 6.7. However, neither does it impose a phylogenetic branching interpretation. Because the distances are not presented in the form of a tree, it becomes less suggestive that a tree structure is the most appropriate. Indeed, this two-dimensional map suggests some correlation with geography, a pattern expected under isolation by distance— the farther two populations are apart geographically, the less gene flow is likely to be shared, and the larger the genetic distance between them.

If geographic distance has influenced rates of gene flow, then we would expect to see a strong correlation between geographic and genetic distance. Luca Cavalli-Sforza and colleagues applied the isolation by distance model to their large database of allele frequencies for classical genetic markers around the world, finding an overall good fit.[20] Elise Eller also looked at isolation by distance in her analysis of 60 microsatellite loci from 15 populations in Africa, Europe, and Asia and found a high correlation between geographic and genetic distance.[21] Given these results, the hypothesis that worldwide genetic distances are a reflection of isolation by distance, resulting from the effect of geographic distance on gene flow, is supported.

6.4.3   The Need to Consider Regional Differences in Population Size

Variation in gene flow produces patterns in which some populations are more genetically similar. To get a pattern in which one population is more divergent (such as in the example in Fig. 6.10), the number of migrants exchanged must be less than among other populations. The genetic divergence of Africa might therefore reflect a smaller number of migrants exchanged with Africa than exchanged between other regions, such as between Europe and Asia. Although such migration models are overly simplistic, treating geographic regions as aggregates, the initial results suggest that African genetic divergence may be a consequence of limited gene flow.

However, this conclusion poses a problem if we also consider within-group variation, as discussed in Chapter 5. If gene flow into Africa is less than for other regions, thus explaining the higher genetic distances, then we would also expect *lower* rates of diversity in Africa, which counters our previous observations. To illustrate this problem, I computed the expected relative heterozygosity for the hypothetical example presented in Fig. 6.9 (the relative heterozygosity is the amount of heterozygosity expected relative to the total heterozygosity if all populations were pooled—see chapter notes for the specific derivation[22]). The expected relative heterozygosity within population A (the most genetically divergent) is 0.852 that of all populations pooled together, which is lower than the expected relative heterozygosity in populations B and C (0.932). Population A has the *lowest* relative heterozygosity rather than the highest. This makes sense because heterozygosity is proportional to gene flow. Populations with more gene flow will have higher levels of heterozygosity, all other things being equal.

We now face a problem—simultaneously explaining the higher genetic divergence of Africa *and* the higher genetic diversity within Africa. If gene flow into Africa is limited, we predict higher divergence and lower diversity. On the other hand, if we allow more gene flow into Africa than other regions, then we predict lower divergence and higher diversity. The problem with both predictions is that our observations show that Africa has higher divergence *and* higher diversity. On the surface, it would seem that the development of a population structure model to explain one finding automatically contradicts the other finding. The simple model used so far can predict either higher African divergence or higher African diversity, but not both.

The solution to this seeming paradox lies in closer examination of the simple model used in Fig. 6.9. The model assumed equal population sizes. What would the impact be if there was variation in population size? Population size enters into this problem in two ways. First, smaller populations will drift more, producing higher genetic distances and lower levels of within-group diversity. Second, the *rate* of gene flow into a population is a function of the number of migrants and the population size. Imagine, for example, that two populations, A and B, exchange one migrant each generation. Imagine further that the population size of A is 2000 and the population size of B is 500. The *rate* of gene flow into population A from population B is equal to $1/2000 = 0.0005$. The rate of gene flow into population B from population A is equal to $1/500 = 0.0020$. When the number of migrants exchanged is symmetric and population sizes are different, then the rate of gene flow is asymmetric. There will be less gene flow into a larger population than out of it.

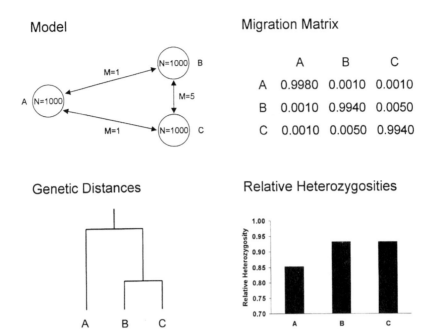

Figure 6.12. *Simulation of genetic distances and heterozygosities for three populations using a model of equal population sizes and unequal rates of gene flow. Populations B and C experience more gene flow between them each generation (0.0050) than either does with population A (0.0010). The genetic distances reflect this variation in gene flow, showing that populations B and C are more genetically similar to each other than either is to population A. Expected relative heterozygosity is lowest in population A, reflecting less incoming gene flow relative to the other two populations. See chapter notes for computational details. This simulation does not fit data from living human populations, in which the most genetically divergent region (Africa) is also the most genetically diverse.*

Figures 6.12–6.14 show the effect of variation in population size on genetic distance and heterozygosity. Each figure starts with the same *number* of migrants as in Fig. 6.9—each generation one migrant is exchanged between populations A and B, one migrant is exchanged between populations A and C, and five migrants are exchanged between populations B and C. Each of these three figures shows the basic migration model, the migration matrix of *rates*, a dendrogram of the genetic distances, and a graph of the within-group relative heterozygosities. Figure 6.12 provides a summary of the hypothetical example used so far. In this example, all three populations have the same size (= 1000). Therefore, the migration *rate* matrix shows that the rate of gene flow into A from B or C is lower than the rate between B and C. As expected, the genetic distance dendrogram shows greater similarity between populations B and C, whereas population A is more distant. The heterozygosity graph shows that population A is the least variable, in this case a consequence of lower rates of incoming gene flow. Overall, this model is not an appropriate one for humans because the most divergent population is also the least variable, whereas we expect (based on Africa) that the most divergent population should be the most variable. Clearly, a model of equal population sizes does not work well here.

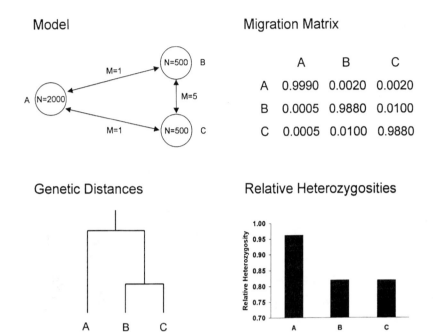

Figure 6.13. *Simulation of genetic distances and heterozygosities for three populations using a model of unequal population sizes in which population size is largest in population A. The greatest amount of gene flow is between populations B and C (0.0100). There is now less gene flow into population A (0.0005) from both B and C than out of it (0.0020 into both B and C). The genetic distances reflect this variation in population size and gene flow, showing that populations B and C are more genetically similar to each other than either is to population A. Expected relative heterozygosity is highest in population A. See chapter notes for computational details. This simulation fits data from living human populations, in which the most genetically divergent region (Africa) is also the most genetically diverse.*

Figure 6.13 summarizes the results when we take the same *number* of migrants between regions as in Fig. 6.12 but set population A to be larger (2000) than either population B or C (both set equal to 500 in this example). As in the previous example, the rate of gene flow between populations B and C is still greater than the rate of gene flow into population A. As a result, the genetic distances to population A are still the highest, giving virtually the same dendrogram as in the previous figure. The rate of gene flow *into* population A from populations B and C (0.0005 each) is now less than the rate of gene flow *out* of population A into populations B and C (0.0020 each). Heterozygosity is highest in population A. Figure 6.14 takes the same *number* of migrants as in the previous two examples but now makes population A the smallest. The same pattern of genetic distances is produced, in which population A is the most divergent, but heterozygosity is lowest in population A.

We can now summarize the results of these three hypothetical examples. In each case, the underlying number of migrants was the same, with more migrants between populations B and C than with population A. The only thing that changed was the distribution of population sizes. We can compare the genetic divergence and heterozygosity of population A for all three cases:

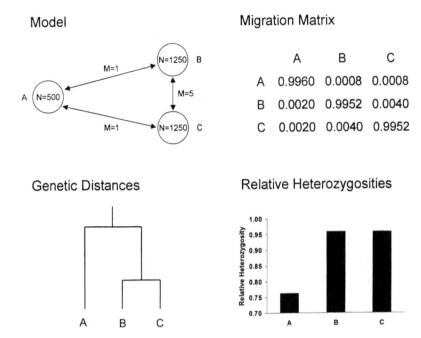

Figure 6.14. *Simulation of genetic distances and heterozygosities for three populations using a model of unequal population sizes in which population size is smallest in population A. The greatest amount of gene flow is between populations B and C (0.0040). There is now more gene flow into population A (0.0020) from both B and C than out of it (0.0008 into both B and C). The genetic distances reflect this variation in population size and gene flow, showing that populations B and C are more genetically similar to each other than either is to population A. Expected relative heterozygosity is lowest in population A. See chapter notes for computational details. This simulation does not fit data from living human populations, in which the most genetically divergent region (Africa) is also the most genetically diverse.*

Model 1: When population sizes are equal, population A shows the greatest divergence and the lowest heterozygosity.

Model 2: When population A is the largest, it shows the greatest divergence and the highest heterozygosity.

Model 3: When population A is the smallest, it shows the greatest divergence and the lowest heterozygosity.

Although each model produces a situation where population A is the most genetically distant, only Model 2, where A has the largest population size, produces higher levels of diversity in population A.

What does this mean in terms of the genetic evidence for modern human origins? So far, we have seen an observed pattern in which Africa has the highest levels of within-region diversity and is also the most genetically different of all major geographic regions. These observations, although often used to support a replacement model of human origins, can also be explained with a population structure model of gene flow and genetic drift, as long as Africa had the largest long-term average population size. As argued in Chapter 5, a larger African population makes sense geographically and ecologically. If true, then two of the observations from the genetic data (greater African divergence and greater African diversity) may simply reflect a larger African population.

6.4.4 Estimating Ancient Migration

There is no question that migration between geographic regions occurs today and has occurred in the past. Both multiregional and replacement models must accommodate gene flow between geographic regions. As with many issues in the modern human origins debate, the argument is over *when* this gene flow occurred. Multiregional models suggest that interregional gene flow has occurred throughout prehistory going all the way back to the first movement of some *Homo erectus* populations out of Africa. Although there might have been times when one or more regions (or populations within regions) were isolated, they were never separated long enough to give rise to a new species. When contact was reestablished, all populations would have continued to be part of a single evolving lineage. Any attempt to model or test a multiregional model must therefore consider migration.

Migration is also necessary when considering regional replacement. Under a replacement model, all of our ancestors arose from Africa 200,000 years ago. As some of our ancestors moved out of Africa, they might have been isolated for some length of time. The "weak Garden of Eden" model developed by Henry Harpending and colleagues suggests that there was regional isolation after the initial dispersal of humans out of Africa 100,000 years ago.[23] Later in time, these isolated groups grew in size and experienced gene flow between them. Even though such models postulate some isolation, they also acknowledge more recent gene flow between regions (within the past 50,000 years or so). Even if a replacement model is correct, it cannot be accurately portrayed by a phylogenetic branching model that allows no gene flow. Regional populations of living humans are not separate species and cannot be modeled as though they were.

Gene flow between geographic regions would have occurred for any origin model, and the only difference is how long it has occurred. Two million years? Fifty thousand years? Sometime in between? Regardless of the answer, another question arises—*how much* gene flow was likely? Living human populations are relatively homogeneous (low F_{ST} values), which can be explained in terms of gene flow. As discussed above, we estimate an average rate of roughly 0.25–0.50 migrants per generation between each pair of geographic regions. To explain larger African genetic distances, we need some variation in those numbers, such that the rate of gene flow into Africa is less than between non-African regions. It would be useful if we could extend these basic inferences and estimate the actual level of migration needed between regions to replicate a given matrix of genetic distances.

My colleague Henry Harpending and I developed such a method.[24] Our method takes a matrix of genetic similarities and estimates of the relative population sizes and computes a matrix of the *number* of migrants needed per generation between each pair of populations to replicate the genetic distances between populations. We applied our method to two data sets both of which included two to four local populations within each of four geographic regions: sub-Saharan Africa, Europe, Australasia, and East Asia. One data set consisted of 57 craniometric measurements

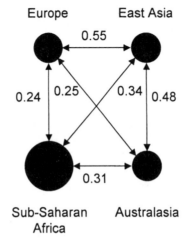

Figure 6.15. *Reconstructed migration matrix between four regional clusters based on pooled data for classical genetic marker loci (93 alleles for 37 loci) and craniometric traits (57 variables). This reconstruction was based on a model in which the total species size is 10,000 reproductive adults with 50% in sub-Saharan Africa and the remainder equally divided among the other three regions (Relethford and Harpending 1995).*

taken from recent human crania, and the other consisted of 93 alleles for 37 classical genetic marker loci. We used a relative population size of 50% for Africa and 17% for each of the other three regions (based on our initial craniometric analysis, as described in Chapter 5).

Figure 6.15 provides a schematic drawing of the results for the distance measures based on the pooled results for both data sets. The numbers indicate the estimated number of migrants exchanged each generation in each direction. The numbers range from 0.24 to 0.55, with an average of 0.36 migrants, which is roughly equivalent to one migrant every three generations. As noted earlier, this migration model is an oversimplification, because it is likely that long-range migration took place sporadically and would have involved *groups* of migrants. The numbers just provide us with relative averages over times that are useful for simple modeling. There are two main results from our analysis. First, migration alone could replicate the observed genetic distances between regions—phylogenetic branching is not required. Second, these estimates suggest that the total number of migrants per generation into Africa (0.89) is less than into any other region. This lower migrant number, combined with the larger population size of Africa, suggests a lower *rate* of gene flow into Africa than into any other region, a situation that results in higher genetic distances to Africa.[25]

We must keep several caveats in mind when interpreting these results. First, the model assumes that the *number* of migrants is symmetric; that is, the same *number* migrate back and forth between any pair of regions. This may not be a major problem, because migration matrix studies show that the symmetric model fits pretty well even when the actual pattern is not symmetric.[26] A second problem is that our model assumes equilibrium, which may not be the case. This, too, is not a major problem because violation of the equilibrium assumption will affect the actual estimates but not the major conclusion that migration can replicate genetic distances.

Perhaps the major problem with our analysis is that it provides evidence that migration alone *could* have produced observed patterns of genetic distances, but it does not *prove* that migration was the sole factor. Although a pure phylogenetic branching model is unrealistic, it is possible that the true population history included *both* population branching *and* migration. Anne Bowcock and colleagues reached a similar conclusion, suggesting that phylogenetic branching out of Africa was obscured by gene flow from East Asia and Africa into Europe.[27] Felsenstein's observation that phylogenetic branching and migration can mimic one another continues to plague the interpretation of genetic distances.

An additional problem is that, given migration at least through the more recent past, any previous history of phylogenetic branching could be "erased" by gene flow. Imagine that the African replacement model is correct and that the pattern of genetic distances 100,000 years ago reflected this history. If there was migration between populations of this new species across the Old World, it is possible that the earlier pattern of genetic distances would be erased, and we would only be seeing the more recent migration history. Because we lack genetic distances from 100,000 years ago (and because the fossil record is not complete enough to apply these same methods to measurement data), we must rely on our observations of genetic distances between living human populations.

How likely is it that migration will erase previous history? This depends on the rate of migration. Alan Rogers and Henry Harpending have developed a simple measure of the speed at which a migration matrix will lead a set of populations to a new equilibrium, thus erasing any previous history. Their estimate, known as the half-life to convergence,[28] can be computed from a migration matrix, and it provides the number of generations needed to go half of the way to equilibrium values. The genetic distances will reach 50% of the equilibrium value in one half-life, 75% in two half-lives, and so on. On the basis of migration matrix we estimated from genetic marker data, Henry and I found a half-life of roughly 1000 generations. This translates to roughly 25,000 years if we use 25-year-long generations. This means that there would be time for four half-lives during the past 100,000 years, which means that almost 94% (50 + 25 + 12.5 + 6.25) of the distance between the old equilibrium (before migration) and the new equilibrium would be reached in this time. Even if we assume the "weak Garden of Eden" model and limit interregional migration to the past 50,000 years, there is still time for two half-lives. None of these estimates is meant to be highly accurate; rather, they serve to illustrate the basic point that migration tends to erase population history if sufficient time has elapsed. If we further factor in the kinds of rapid demographic changes and increased migration likely since the origin of agriculture, we might expect that genetic distances between populations are not likely to be accurate reflections of very ancient population history. This does not mean that we cannot recover this history, but rather that any attempts to do so must avoid simplistic interpretations and must control for migration.

6.5 SUMMARY

Although regional genetic differences exist in living humans, we are comparatively a homogeneous species (at least when compared to a number of other mammalian species). Variation exists, but it is relatively minor *between* human groups and much larger *within* groups. These figures are even more impressive when we consider the geographic range of living humans—the relative amount of variation between geographic regions is less than found between geographic neighboring populations in other species. This observation suggests two overlapping possibilities—the relatively low values of F_{ST} reflect recent separation of human "races" and/or migration between regions. The first possibility supports an African replacement model, whereas the second could support either a replacement or multiregional model. The values of F_{ST} cannot tell us which is correct.

Studies of F_{ST} tell us about the magnitude of population differences. Studies of genetic distance are used to tell us about the pattern of such differences. These studies are consistent and clear—sub-Saharan Africa is the most genetically distant region. Again, there are several different possible interpretations. Some argue that the finding of higher African divergence fits a replacement model—lower genetic distances between non-African populations is consistent with a more recent splitting of these populations from an initial population moving out of Africa 100,000 years or so ago. However, if the replacement model is correct, it is unreasonable to assume that these regional groups have remained isolated for this entire time. If there is some migration between regions, it will act to reduce group differences and to erase previous genetic history. To further complicate matters, it is possible that the pattern of genetic distances can be explained solely by an ancient pattern of interregional migration, perhaps dating back to the initial dispersal of some *Homo erectus* populations from Africa.

By themselves, studies of genetic distance cannot rule out one origin model over the other. As is the case with coalescent dates (Chapter 4) and levels of within-group diversity (Chapter 5), there are alternative explanations that are equally reasonable. The fact that genetic distances, as well as other genetic information, are compatible with the African replacement model does not prove that it is the correct model.

One common finding across this chapter and Chapters 4 and 5 is the extent to which population size is an important parameter in explaining genetic variation. We have seen how population size affects coalescent dates, levels of within-group diversity, and rates of migration. Chapter 7 also considers population size, but from a different perspective that *may* ultimately provide us with the final answer to the modern human origins debate.

Chapter 7

How Many Ancestors?

A common theme in Chapters 4, 5, and 6 is the importance of population size in evolution. The size of a population affects the level of genetic drift and therefore the level of diversity. Small populations are more likely to lose alleles through genetic drift and will consequently have less genetic diversity. Larger populations experience less genetic drift and will have more genetic diversity. A related function of population size is coalescent dates; the smaller the population, the more recent (on average) the coalescence to the most recent common ancestor. Population size also affects rates of gene flow, as shown in Chapter 5.

The size of a population or an entire species is a reflection of its history, particularly the balance between births and deaths over time. Population sizes fluctuate over time depending on environmental circumstances. When considering human history and prehistory we must also deal with changes in technology and behavior that have affected overall species size. Ecologists look at population size in terms of a population's carrying capacity, the maximum number of individuals that can be supported in a given environment. Throughout hominid evolution the carrying capacity of the species was likely to have been small, such that there were probably no more than five to ten million people living some 10,000 years ago. Ecologically based estimates suggest even fewer before this time—perhaps a million or fewer during much of the Stone Age.[1] The situation has changed dramatically since the spread of agriculture over the past 12,000 years. At present, our species numbers more than six *billion* people and is likely to continue to grow throughout much of the twenty-first century. This is a dramatic increase—roughly 1000-fold in 10,000 years! New genetic evidence, discussed later in this chapter, suggests that the rapid increase in population size began even earlier, perhaps 50,000 years ago.

The main point here is that a species' history is reflected in its changing demography. Traditionally, demographic history has been investigated using ecological and archaeological data, documenting changing patterns of population growth and increased population density over the past two million years. In the past decade, genetic data have also been brought to bear on the question of ancient demography, focusing on two related questions: 1. How many people lived in the past? 2. How did these numbers change over time? Did they increase, decrease, or both? Because patterns of genetic variation are a reflection of demographic history, we have the opportunity to discover much about our species' history by estimating demographic parameters from genetic data. In addition, knowledge of our demographic history may shed light on the modern human origins debate.

As we have seen, much of the genetic evidence can be interpreted in terms of replacement and multiregional models. A number of researchers have moved away from such arguments and looked more directly at demographic history. Several scholars, such as Henry Harpending and Alan Rogers, have argued that the demographic history revealed from genetics provides us with indirect support of a replacement model. They argue that the average size of our species over time as estimated from genetic data is too small to be feasible under a multiregional model. The genetic evidence suggests no more than 10,000 reproductive-aged adults throughout much of our species' history. If true, then it is difficult to imagine how so few people could have been spread out over several continents and remained connected by gene flow as required by the multiregional model. A more plausible explanation, in their opinion, is that the small number reflects the fact that our ancestors 200,000 years ago lived in a small section of Africa and were a separate species from the archaic humans living elsewhere. Their interpretation is that the estimates of species' size support a replacement model.[2] In my view, this finding is the strongest genetic evidence for replacement. However, there are some unanswered questions, primarily dealing with the relationship between actual population numbers and the genetic estimates. It is possible that the genetic estimates are quite a bit lower than the actual population numbers if certain models of population dynamics apply to ancient human populations.

7.1 DIFFERENT CONCEPTS OF POPULATION SIZE

The size of a population seems to be a straightforward concept, but it can be defined in different ways, each having different interpretations for human evolution. The simplest measure of population size is the *census population size* (given the symbol N_c), which is the total number of all individuals. This type of enumeration is done routinely today. For example, the United States conducts a national census every 10 years in an attempt to count all residents. Because we cannot be sure of having counted everyone (not to mention that fact that people are born and die during the time it takes to conduct a census), our final numbers are estimates. Statistical methods are used to adjust numbers for possible underreporting and other potential problems, but the initial data are a summary count of all people in a household. We obviously cannot do this directly for the ancient past, so prehistoric census population size estimates are derived in a different manner—estimates of usable land area are multiplied by estimates of population density (often based on living hunting and gathering societies) to get a rough estimate.

Census population size is a useful measure for a variety of demographic and ecological analyses but is not immediately useful for predictions of genetic variation. The problem is that census population size refers to *all* members of a population, whereas genetic models deal only with those individuals that reproduce in any given generation. We refer to this latter number as the *breeding population size* (N_b). Ideally, we would count up all individuals that reproduce in a given generation and not count any that are too young, too old, or for one reason or another did not reproduce.

Such a detailed estimate requires information on each person's reproductive status. Most often, we use an approximation to this number by considering the proportion of the population that is of reproductive age. Any population is made up of individuals of different ages, ranging from newborns to the elderly. For our purposes here, we can divide a population into three groups: 1. Prereproductive, 2. Reproductive, and 3. Postreproductive. Because genetic models focus on what occurs within a given generation, we only count those of reproductive age at one point in time to estimate the breeding population size. The other age groups refer to different generations. Prereproductive refers to people who have not yet reproduced because they are too young. Many will eventually reproduce, but we would then count them in the next generation. Postreproductive refers to people who have already reproduced and are now older than reproductive age and would already have been counted in the previous generation. On the basis of the age structure of human populations, we often suggest that the breeding size is roughly one-half in hunting and gathering societies and roughly one-third in industrialized societies, where more people live to older ages.[3] This is a very crude approximation, and we must take into account the possibility for a lot of variation about this figure. Still, these crude ratios are good for some rough order of magnitude comparisons.

Even with accurate information on breeding population size, we cannot simply plug these figures into population genetic models. As is the case with most mathematical models, population genetics makes use of some simplifying assumptions. When these models incorporate breeding population size they often assume equal numbers of males and females, monogamy, lack of inbreeding, no differentiation of groups, and constancy in population size over time, among others. It is quite clear that some (or all) of these assumptions are violated in many human populations. This is particularly a problem for long-term considerations of human demographic history, where the assumption of constant population size, generation after generation, is unrealistic. We face a common problem in the development of any mathematical model—the assumptions are necessary to make the mathematics tractable but the imposition of potentially unrealistic assumptions make any results questionable.

Fortunately, there is a way around this difficulty when dealing with the genetic effects of population size—the development of the concept of *effective population size* (N_e). The effective size of a population is an estimate of the breeding size that would apply if all of the assumptions applied. Various formulae have been developed to derive estimates of effective size from breeding population size.[4] A simple example helps show the relationship between breeding population size and effective population size. One of the assumptions of our models is that there are an equal number of reproductive-aged males and females. What happens when this is not the case? Consider a hypothetical population that has 30 adult males and 270 adult females. The breeding size of this population is simply the total of reproductive-aged males and females, or 30 + 270 = 300. However, if we actually used the estimate of 300 in any predictive model (say genetic drift or inbreeding), we would actually be overestimating the true genetic size, because these models assume an equal number of adult males and females. How can we get to the true genetic size from these numbers? The formula for effective population size for unequal numbers of males and females is simple:

$$N_e = \frac{4MF}{M+F}$$

where M is the number of adult males and F is the number of adult females. Plugging our numbers of $M = 30$ and $F = 270$ into this formula gives

$$N_e = \frac{4MF}{M+F} = \frac{4(30)(270)}{50+250} = \frac{32400}{300} = 108$$

This means that the hypothetical population of 30 males and 270 females will show the same amount of genetic drift (or inbreeding) that would be found in a population of 108 adults where there were equal numbers of males and females (54 each in this case). In this example, we see that major deviations from the assumption of equal numbers of males and females can result in a substantially reduced effective population size. This would not be the case if the numbers of males and females were more similar. For example, if there were 140 males and 160 females, the effective size (299) is virtually the same as the breeding size (300).

A factor that is more likely to have affected human effective population size is temporal change. Many models of population genetics assume that the breeding population size has been constant over time. When this is not the case, and there are major changes in breeding population size over time, the average effective population size can be reduced quite a bit. As an example, consider the level of inbreeding in a hypothetical population that starts with a breeding size of 100 and doubles in size each generation for four more generations. The population sizes for the six generations are 100, 200, 400, 800, and 1600. To make matters simple, we assume zero inbreeding to begin with and no migration into the population. The level of expected inbreeding in this case is a function of population size—smaller populations are likely to show more inbreeding because there are fewer available mates that are not related. Using a simple model (see chapter notes), we predict that the level of inbreeding after five generations is 0.0097 (this number refers to the probability that two individuals will share the same gene because of inheritance from a common ancestor). Although in this example it was easy enough to use different values of breeding size in each generation, most mathematical models work better when population size is the same from one generation to the next. Although this is clearly not the case in this example, the use of effective population size allows us to derive a single number that will produce the same inbreeding after five generations. When breeding population size changes, we can compute effective population size as the harmonic mean, defined as

$$N_e = \frac{t}{\sum (1/N_i)}$$

where N_i refers to the breeding population size in generation i and there are t generations. In this example $t = 5$ and we compute the effective size as

$$N_e = \frac{5}{(1/100)+(1/200)+(1/400)+(1/800)+(1/1600)} = 258$$

The amount of inbreeding expected in a hypothetical population with $N_e = 258$ is equal to 0.0097—the same value as expected under the model of changing population size. The amount of inbreeding expected after five generations of population growth[5] is the same as the amount of inbreeding expected under a model of constant breeding size of 258 adults. Thus the model of constant effective population size (258, 258, 258, 258, 258) produces the same end result as the model of population growth (100, 200, 400, 800, 1600). Because constant population size is more tractable mathematically in many situations, the concept of effective population size allows us simplicity in modeling by adjusting reality to the assumptions of our models.

The two examples given above have focused on two different genetic outcomes—genetic drift in the first and inbreeding in the second. In many cases, the same effective population size formula can be used. In some cases, however, the two measures may be somewhat different. To distinguish them, population geneticists often refer to *variance effective size* (relating to the variance in allele frequencies expected under genetic drift) and *inbreeding effective size* (relating to the level of consanguinity expected between any two individuals). Because many of our concerns here are related to coalescence theory and the genetic relationship between individuals, all discussions of effective population size refer to inbreeding effective size unless otherwise indicated.

To summarize, it is useful to distinguish between three different measures of population size—census population size, breeding population size, and effective population size. Although census population size can be estimated from ecological and archaeological inference, genetic estimates refer to effective population size. A major debate in modern human origins is the exact relationship between these measures and how to make inferences from one based on another. If, for example, we estimate effective population size as equal to 10,000, what can we say about breeding population size? Sometimes effective size (N_e) is less than breeding size (N_b), but sometimes it might be equal or even greater. To relate either of these numbers to census population size (N_c) we need to make assumptions about the age structure of a population. We now turn to various estimates of population size based on ecological and genetic data to address some of these problems.

7.2   ESTIMATING ANCIENT POPULATION SIZE

Various methods have been derived for estimating prehistoric population size. In some cases, archaeological data can be used to derive population size estimates from variables such as the total area of a site, the number and size of houses or other dwelling units, artifact density, and evidence of food consumption, among others.

Such methods often require additional information, such as population density, that is derived from ethnographic analogy—the study of living hunting-gathering populations to generate ecological and demographic inferences.

7.2.1 Ecological Estimates of Census Size

The total census size of our ancestors is more often based on ecological information on total land area used combined with data on population density derived from ethnographic analogy. In the simplest terms, we can estimate total census size (N_c) if we have an estimate of total land area (A) and population density (D). The total census size is then simply A times D ($N_c = AD$). The problem is deriving reasonable estimates of usable land area and population density, because such numbers would have varied across different environments. Some environments can support higher population densities; as mentioned in Chapter 5, savanna environments can support higher population densities than desert environments. Archaeological evidence also shows that these numbers have increased over time. As human ancestors spread from Africa and continued to adapt to new environments, the total usable land area increased. Continued cultural adaptations most likely also led to increases in population density in any given environment. The combined increases in land area and density both contributed to an increase in population.

Fekri Hassan has looked at changes in land area and population density over time, combining inferences from archaeological data and ethnographic analogy with ecological data on biomass and likely productivity under a hunting and gathering way of life. Hassan derived global estimates[6] of the total occupied land area, population density, and total population size for our ancestors over the past 2.5 million years broken down into different stages of the Paleolithic[7] (literally "old stone age"). The Basal Paleolithic is the time of the first stone tools roughly 2.5–2.0 million years ago. The Lower Paleolithic refers to the stone tool culture of *Homo erectus* and early archaic humans, roughly 2.0 million to 250,000 years ago. The Middle Paleolithic refers to the stone tool cultures of Neandertals and other humans between 250,000 and 40,000 years ago. The Upper Paleolithic refers to the stone tool cultures usually associated with modern humans in Europe between roughly 40,000 and 12,000 years ago.

Hassan's estimates are shown in Fig. 7.1. Total land area is estimated to have increased from roughly 20 million square kilometers at the beginning of the Paleolithic to 60 million square kilometers by the time of the Upper Paleolithic. Total land area is estimated to have increased from roughly 20 million square kilometers at the beginning of the Paleolithic to 60 million square kilometers by the time of the Upper Paleolithic. Population density is estimated to have been rather low during much of the Paleolithic, averaging about 0.025 persons per square kilometer, and then increasing to 0.1 persons per square kilometer during the past 40,000 years. The estimated total population size is estimated to have ranged from 400,000 roughly two million years ago to six million during the Upper Paleolithic. Because most of the estimates of effective population size from genetic data (see below) relate to the time period marked by coalescence, the most relevant figures from Hassan's analysis are for the Lower and Middle Paleolithic. Hassan estimates a total world population of 800,000 during the Lower Paleolithic and 1.2 million during the Middle Paleolithic, for an average of one million.

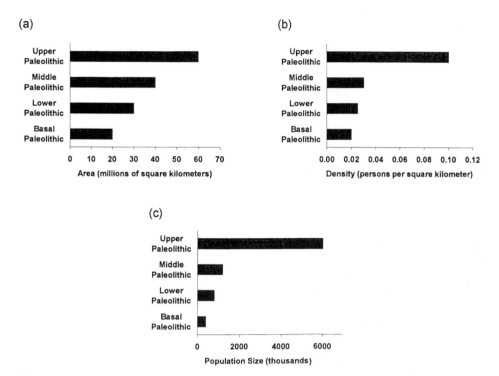

Figure 7.1. *Estimates of prehistoric demography from Hassan (1981, p. 199). a. Occupied land area. b. Population density. c. Population size.*

Other researchers have obtained roughly similar estimates of total census size. Hassan reviewed other estimates of Lower Paleolithic population size, which range from 125,000 to 400,000, and Middle Paleolithic population size, which range from 1.0 to 1.3 million. Henry Harpending and colleagues obtained a much lower figure of 125,000 for the Middle Paleolithic, but their estimate was a minimum possible estimate.[8] Overall, these studies suggest that ancient census size ranged from several hundred thousand to one million or so hominids.

7.2.2 Genetic Estimates of Species Effective Size

Estimates of population size have also been obtained from information on the degree of genetic variation that exists in living human populations. As noted earlier, a species' genetic diversity is a function of mutation rate, elapsed time, and population size. If we make the assumptions that the genetic traits we survey are neutral (no natural selection) and equilibrium has been reached, then, given an estimate of the mutation rate, we can estimate the average long-term effective population size. The exact method of estimating species effective size depends on the specific traits analyzed (e.g., classical genetic polymorphisms, DNA sequences, etc.).

One example of this approach is found in a comparative study by Masatoshi Nei and Dan Grauer, who estimated effective species size for a variety of organisms based on heterozygosity of protein-coding loci.[9] The evolution of neutral genes is expected to result in a level of heterozygosity at equilibrium, H, defined as

$$H = \frac{4N_e\mu}{1+4N_e\mu}$$

where N_e is the species effective size and μ is the mutation rate per generation. This equation can be rewritten to predict effective population size from heterozygosity and mutation rate, as

$$N_e = \frac{H}{4\mu(1-H)}$$

Nei and Grauer used this method to estimate the species effective size of the human species based on a heterozygosity estimate of $H = 0.143$ and an average mutation rate per generation μ of $= 0.000003$, which gives an estimate of N_e of roughly 14,000.

Similar methods have been developed to estimate effective population size from various measures of nuclear and mitochondrial DNA sequence diversity. In general, these methods rely on the fact that expected diversity is proportional to effective population size (times mutation rate). If we have an estimate of the mutation rate per generation for a given DNA sequence, we can estimate effective population size. Effective size can also be estimated from coalescent dates because, as shown in Chapter 4, coalescent dates are a function of population size.

Recent estimates of the long-term effective size of our species are similar and consistently suggest a value of roughly 10,000. Naoyuki Takahata revised the classic work of Nei and Grauer, using updated mutation rates and adding estimates from nuclear and mitochondrial DNA sequence diversity.[10] His estimates of effective species size range from 9,000 to 12,000 based on classical genetic marker loci and from 8,000 to 11,000 based on nuclear DNA sequences. Takahata's estimates from mitochondrial DNA range from 3,500 to 4,600. Keep in mind, however, that estimates from mitochondrial DNA refer to the effective *female* population size. If we assume equal number of males and females, we double this figure to get an estimate of total effective species size of 7,000–9,400.

All of these estimates suggest a rough order of magnitude estimate of a total effective species size of 10,000. As with coalescent dates, there is often large confidence intervals, so that some researchers, such as Alan Templeton, suggest that the *actual* effective species size might be as high as 200,000–500,000.[11] Although this is possible, it is also worth noting that the same general figure of 10,000 has been found across many traits and analyses, suggesting some confidence as an order of magnitude estimate.

Some loci provide slightly different estimates. Steve Sherry and colleagues estimated N_e as roughly 18,000 based on *Alu* insertion polymorphisms,[12] and Francisco Ayala obtained an estimate of roughly 100,000 based on the *DRB1* gene for the major histocompatibility complex.[13] Some of this variation has to do with the time period considered. The genetic estimates provide an idea of the average effective size over time dating back from the present to the date of coalescence. Estimates of effective size from mitochondrial DNA are roughly 10,000 (double the number of females) and correspond to a coalescence date of roughly 200,000, suggesting that the average effective species size has been 10,000 over the past 200,000 years. Y chromosome polymorphisms provide similar estimates. Current evidence suggests that effective size was also relatively low over the past 1.5 million years. Rosalind Harding and colleagues' analysis of the beta-globin gene suggests the same effective size back over the past 800,000 years.[14] Steve Sherry and colleagues obtained a slightly higher estimate ($N_e = 18,000$) over the past 1.4 million years. Ayala's larger estimate of 100,000 applies to a time period of over 60 million years, so this estimate includes earlier primate ancestors.

Overall, the genetic data are consistent with the view that the effective population size of our ancestors was roughly 10,000 over the past million years or more.[15] Although more data are needed to sample different time periods, the overall conclusion of a small effective species size is well confirmed. The controversy results from the interpretation of these results. At first glance, there appears to be a major discrepancy between ecological estimates of ancient population size of roughly 500,000 to one million and the genetic estimates of 10,000. These two estimates are not strictly comparable, because the ecological estimates refer to total census size and the genetic estimate approximates breeding population size based on a set of assumptions.

What breeding size is implied by an effective population size of 10,000? If we assume that our ancestors had an age structure similar to living hunting-gathering societies, we would expect that the breeding population size would be roughly half the census size. If we take the effective size estimate of 10,000 as an approximation of the breeding population size, this means that the total census size estimated from genetic data is $10,000 \times 2 = 20,000$. Even if the ratio was as low as 1/5, as suggested from Alan Templeton's review of large-bodied mammals,[16] the total census size would still be only $10,000 \times 5 = 50,000$.

We are still left with two very different estimates of average census size, with the ecological estimates ranging from several hundred thousand to a million or more individuals, whereas the genetic estimates suggest only several tens of thousands of individuals. Even after we acknowledge uncertainty in various estimates, the fact that there is an order of magnitude difference between these estimates suggests something is wrong. One interpretation is that the difference between ecological and genetic estimates is indirect proof of an African replacement model. The basic argument, considered in detail below, is that the genetic estimates suggest that our ancestors were too few in numbers to have been spread out very far across the Old World. Instead, these ancestors lived in a relatively small region, which other evidence suggests was Africa. The small number of ancestors is considered a reflection of the birth of a new species in Africa, followed by replacement of other hominid populations throughout the rest of the Old World.

Before considering this argument in detail, as well as possible alternatives, we need more information on the demographic history of our ancestors. As used above, the genetic estimates of effective species size are long-term averages. An estimate of $N_e = 10,000$ only tells us so much. It does not tell us anything about the specific demographic history. There are many different scenarios that can give rise to an average population size of 10,000, including cases of population growth, decline, and stability. Our interpretations of low effective species size are contingent to some extent on some knowledge of ancient demographic trends. Before considering the full implications of low species effective size, we need to consider the genetic evidence on past patterns of demographic change.

7.3   A PLEISTOCENE POPULATION EXPLOSION?

A number of genetic studies suggest that our ancestors had undergone a rapid population explosion over the past 50,000–100,000 years. This suggestion has implications for interpreting the fossil and archaeological records as well as providing greater insight into the evolutionary significance of low effective population size. Evidence for a prehistoric population explosion is apparent from analysis of gene trees and other methods, many of which show a characteristic pattern of population growth. Other methods, such as those developed by Alan Rogers and Henry Harpending, provide a means by which to estimate the timing of this growth.

7.3.1   Gene Trees and Population Expansion

Reconstruction of gene trees has provided us with strong evidence for a past expansion in population size (as used here and in the literature, "expansion" refers to a change in population size but not necessarily a geographic expansion). Simulations of gene coalescence, comparing the model in which population size remains constant with the model of population growth, help to illustrate this finding.[17] Such simulations consider the probability of coalescence in any generation as a function of population size; as described in Chapter 4, coalescence is more rapid in a small population than in a large one.

Before considering the results of several simulations, it is useful to know that coalescent simulations use somewhat different measures of effective population size and the timing of population expansion. As noted earlier, mtDNA diversity is directly proportional to effective female population size and the aggregate mutation rate as

$$\theta = 2N_f\mu$$

For simulation (and certain analytic) purposes, it is easier to use θ directly as a measure of population size (which it is if you consider population size in terms of $1/2\mu$ individuals).

Figure 7.2 presents four different simulations of mitochondrial DNA coalescence among 10 DNA sequences in a population of constant size based on a constant value of $\theta = 5$. Note that each tree is different, as expected from a random process. Some trees coalesce earlier, such as the one in the bottom right of the figure. Others coalesce later, such as the one in the upper right of the figure. This variability suggests caution in interpreting coalescent dates; because in practice we reconstruct actual trees from a single locus (such as mitochondrial DNA), we have to always consider statistical variation. Apart from the variability across these four simulations, there is a consistent pattern; in each case, most of the coalescent events occur close to the present, with only one or two older coalescent events.

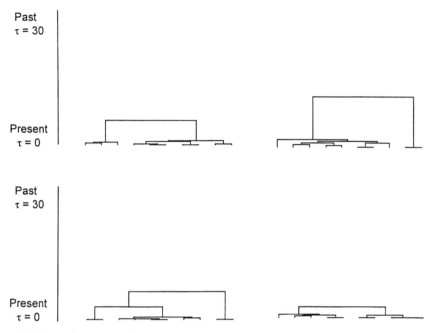

Figure 7.2. *Four simulations of mitochondrial DNA coalescence among 10 DNA sequences in a population of constant size where mtDNA diversity is θ = 5. The time axis is scaled to be the same as in Fig. 7.3.*

Simulated gene trees under a model of population expansion are quite different. These simulations use the "sudden expansion" model in which population size increases "instantaneously" at t generations in the past. It is useful to express the time of expansion in mutation units as

$$\tau = 2\mu t$$

where the timing of the expansion is measured in terms of $1/2\mu$ generations. Figure 7.3 shows the results of four separate simulations under a model in which the population size increased 1000-fold halfway along the time axis. This was modeled by increasing mtDNA diversity from 5 to 5,000 (because the population size increased by a factor of 1000). This expansion was modeled to occur $\tau = 15$ units in the past. The vertical scale was set to show a maximum of $\tau = 30$ units. All of the trees show much "deeper" roots, where all of the coalescent events occur far back in time. Such trees are generally referred to as "star shaped" because if one draws the tree in the form of a circle (as in Fig. 4.6), the lines look like light rays radiating out from a star. Some have suggested (and I prefer) the term "comb shaped" because the lines on the tree clearly resemble the teeth of a comb. This shape results from the changes in population size. There will be relatively few coalescent events after population expansion because the population is so large and the probability of expansion is inversely related to population size. Instead, most of the coalescent events will have taken place immediately before population expansion when the population size was small. Comparing Figs. 7.2 and 7.3, we see that the two demo-

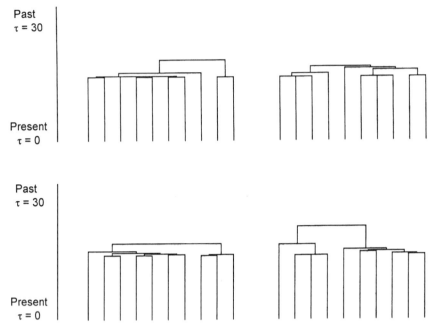

Figure 7.3. *Four simulations of mitochondrial DNA coalescence among 10 DNA sequences under population growth. The population begins with mtDNA diversity of θ = 5 and increases in size 1000-fold at time τ = 15 generations (expressed in mutational time units). The time axis is scaled for a maximum of 2τ generations.*

graphic models (constant size and population expansion) generate different expectations. Because most gene tree analyses that have been conducted show deep lineages and the comb-shaped pattern, a past population expansion is supported.

7.3.2   Mismatch Distributions and Frequency Spectra

Evidence of past population expansion is also evident in two other graphic methods known as the mismatch distribution and the frequency spectrum. Mismatch distributions are based on the pairwise comparison of all DNA sequences in a sample. As described in Chapter 4, each individual sequence is compared with all other sequences, counting the number of nucleotide differences between each pair. Imagine, for example, that there are DNA sequences from four individuals, labeled A–D. We would compare sequence A with sequence B and count the number of nucleotide differences. We would then compare sequence A with sequence C, and A with D, and keep track in each case of the number of differences between each pairing. We would then compare B with C, B with D, and C with D, so that all possible pairs are compared. If we have *n* sequences we will wind up making $n(n - 1)/2$ pairwise comparisons. The final step is to generate the mismatch distribution, which is simply a histogram of the number of differences.

The frequency spectrum provides another way of looking at sequence differences.[18] Here, we simply plot the proportion of sites that occur at a given frequency as a function of the frequency in the sample. We start by examining two alternative nucleotides at a given site in the DNA sequence and count how many times each occurs. For example, if we examine a particular site in a sample of 30 people, there are a number of possible counts—one nucleotide could occur in one person but not in the other 29, one nucleotide could occur in two people but not in the other 28, and so on. Because we do not usually know which nucleotide was the ancestral form, we cannot tell which variant was the mutation. The method of frequency spectrum analysis starts by "folding" the distribution at one-half of the sample size, so that variants occurring in one out of 30 people are counted the same as those occurring in 29 out of 30 people. Likewise, a variant that occurs in 5 out of 30 people is counted the same as one that occurs in 25 out of 30 people. We then go through all sites, tallying the number of times a variant occurs. If, for example, the variant occurs 1 out of 30 times (or 29 out of 30), we would tally up one observation for the value of "1." Likewise, if we found that a variant occurred 5 times out of 30 (or 25 out of 30), we would tally up one observation for the value of "5." We continue building the histogram site by site. The height of the final histogram represents the proportion of times sites occur at a given frequency in the sample.

As with gene trees, mismatch distributions and frequency spectra are different under the contrasting demographic models of constant population size and population expansion. Figure 7.4 shows the results of a simulation of 30 mitochon-

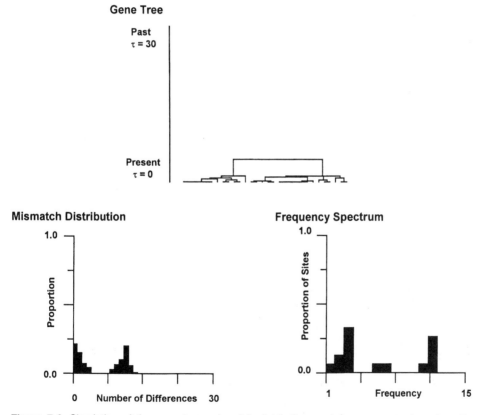

Figure 7.4. Simulation of the gene tree, mismatch distribution, and frequency spectrum for mitochondrial DNA coalescence among 30 DNA sequences in a population of constant size where mtDNA diversity is $\theta = 5$. The time axis is scaled to be the same as in Fig. 7.5.

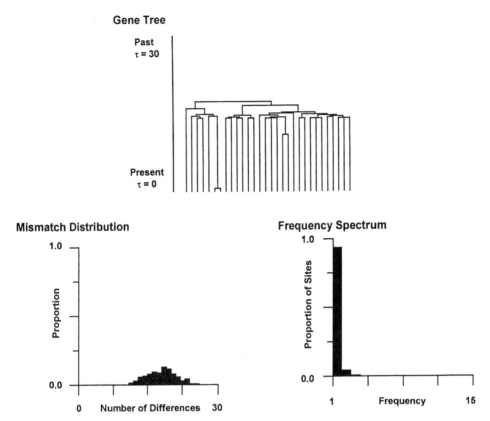

Figure 7.5. Simulation of the gene tree, mismatch distribution, and frequency spectrum for mitochondrial DNA coalescence among 30 DNA sequences under population growth. The population begins with mtDNA diversity of $\theta = 5$ and increases in size 1000-fold at time $\tau = 15$ generations (expressed in mutational time units). The time axis is scaled for a maximum of 2τ generations.

drial DNA sequences under a model of a constant population size ($\theta = 5$). The gene tree resembles those in Fig. 7.2, with short branches and most coalescent events occurring a relatively short time ago in the past. The mismatch distribution and frequency spectrum are bimodal. These graphs look much different when considering a model of population growth, as shown in Fig. 7.5, which is based on a simulation of 30 mitochondrial DNA sequences in a population that expanded 1000-fold at time $\tau = 15$. Here, the gene tree shows deep branches and most coalescent events occurred before population expansion when the population size was low. The mismatch distribution now has a single mode and is roughly bell shaped. The frequency spectrum is heavily skewed to the right, with almost all variant sites occurring only once.

Given these different expected outcomes for constant population size (Fig. 7.4) and population expansion (Fig. 7.5), we can now examine some *actual* DNA sequences to determine which model provides the best fit to reality.[19] Figure 7.6 is the mismatch distribution based on a world sample of 636 mitochondrial DNA sequences (for 411 sites of the first hypervariable segment of mtDNA). This distribution clearly

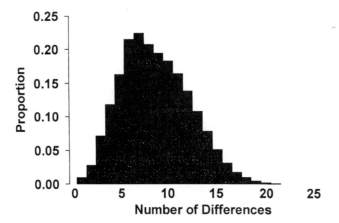

Figure 7.6. *Mismatch distribution for 636 mitochondrial DNA sequences for 411 sites of the first hypervariable region (HVRI) (Harpending et al. 1998) (Data provided by Lynn Jorde).*

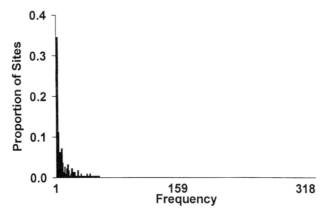

Figure 7.7. *Frequency spectrum for 636 mitochondrial DNA sequences for 411 sites of the first hypervariable region (HVRI) (Harpending et al. 1998) (Data provided by Lynn Jorde).*

has a single mode and closely resembles the expected distribution under a model of population expansion (see Fig. 7.5 for comparison). Figure 7.7 shows the frequency spectrum for the same data. The distribution is heavily skewed to the right, and the vast majority of variants occur only once, as expected from a model of population expansion (see Fig. 7.5 for comparison). As with gene trees, mismatch distributions and frequency spectra based on mitochondrial DNA support ancient population expansion. Henry Harpending and colleagues have presented similar conclusions based on Y chromosome data and *Alu* insertion polymorphisms.[20]

7.3.3 Evidence of Expansion from Microsatellite DNA

Evidence of ancient population expansion has been found using several new methods for analyzing microsatellite DNA. These methods rely on the comparison of various measures of DNA variation with values expected under different demographic scenarios. In each case, the observed patterns of DNA variation fit more closely the values expected under a model of rapid ancient population expansion.

Anna Di Rienzo and colleagues focused on measuring the variance in the number of DNA repeats.[21] Their theoretical work shows that the expected variance under a model of constant effective population size is proportional to the product of mutation rate and population size. Under a model of rapid population expansion, the expected variance is proportional to mutation rate and the time elapsed since expansion. Both models predict a linear relationship between mutation rate and variance in the number of DNA repeats. The two models differ, however, in the expected amount of variation around the expected values, that is, how well the underlying model relating mutation rate and variance in repeat number fits. In general, the model of constant population size will produce more deviations from the expected model, because we expect more genetic drift under a model of constant size rather than one in which the population increased in size (thus reducing the amount of genetic drift).

Di Rienzo and colleagues applied this method to the analysis of 24 microsatellite DNA loci in three human populations (Luos from Kenya, Kaingang from Brazil, and Sardinians from Europe). They examined several mutation models, including the case in which the mutation rate was constant across loci and cases in which there was moderate and high variability in mutation rates across loci. In each case, the observed variance in DNA repeats was more similar to the amount predicted under rapid population expansion than to the amount predicted under the assumption of constant population size. For example, the observed variance in repeat number was 0.81, pooling data from all three populations. The variance expected under constant population size and mutation rate was 2.59, whereas the variance expected under rapid population growth was 0.08. The observed value is clearly more similar to the value expected under population expansion than to the value expected under constant population size. Additional analyses, allowing different parameters for mutation rate, produced the same results, supporting an ancient population expansion.

David Reich and David Goldstein applied two somewhat different approaches to their analysis of 60 microsatellite DNA loci (30 of which were from loci in which two nucleotides repeated and 30 from loci in which four nucleotides repeated).[22] They first noted that the distribution of the number of DNA repeats is expected to be different in cases of constant size versus cases of population expansion; the former tends to produce ragged distributions with multiple peaks, whereas the latter tends to produce smooth distributions with a single peak. Accordingly, they devised a statistical measure that measured the degree of "raggedness" in observed distributions. Their second approach was similar to that developed by Di Rienzo et al., focusing on the fact that constant population size would produce more variation in the variance of repeats than a model of rapid population expansion. Reich and Goldstein applied their methods to data from populations across the world and found that both methods showed evidence for an ancient population expansion in Africa but not outside of Africa.

A third study of microsatellite DNA was reported by Marek Kimmel and colleagues, using data from 60 loci from African, Asian, and European populations.[23] Their approach was based on two different ways to estimate θ, the product of mutation rate and effective population size. One way of estimating θ is based on the variance in repeat numbers, and the other way is based on overall homozygosity. Under a model of constant population size, these two numbers are expected to be the same. Under a model of population expansion, however, the variance-based estimate will be larger than the homozygosity-based measure, and the ratio of these two estimates will be greater than 1. Their analyses showed that the ratio was greater than 1 for African (1.12), Asian (1.82), and European (1.33) populations, supporting the hypothesis that population expansion had occurred in all three regions. Their results differ somewhat from those of Reich and Goldstein, who found evidence of expansion only in African populations.

These analyses of microsatellite DNA are in agreement with results from mitochondrial DNA and Y chromosome data, and all point to an ancient increase in the population size of our species. Although many researchers now accept the net results from all of these studies as firm support for ancient population expansion, not everyone agrees. Some, like John Hawks and colleagues, suggest that the evidence for expansion from mitochondrial DNA is actually showing us the expansion of particular DNA sequences rather than the expansion of the human species, a pattern we might expect from natural selection.[24] As noted in Chapter 4, some researchers have suggested that portions of the mitochondrial DNA may have been subject to natural selection. Because a person's mitochondrial DNA is inherited as a unit, equivalent to a single locus, this means that even a small amount of selection for part of the mitochondrial genome might produce what is termed a "selective sweep" that mimics the pattern expected from population expansion. Instead of evidence for an expansion in the size of our species, we might be seeing evidence for the genetic spread of favorable genes, which does not necessarily imply an increase in species size. The same argument could presumably be made for Y chromosome polymorphisms.

The argument that genetic data support an ancient expansion is strengthened by the fact that, in general, studies of microsatellite DNA show the same thing. It is highly unlikely that natural selection for a section of mitochondrial DNA would have anything in common with selection across a range of microsatellite loci. However, Hawks and colleagues note that the evidence from microsatellite DNA might not be that clear, because some studies show somewhat different conclusions. As noted above, the study by Reich and Goldstein did not find evidence for expansion outside of Africa (although they did find expansion within Africa), whereas Kimmel and colleagues' study showed expansion in both African and non-African populations. It is not clear at present to what extent such differences reflect sampling, robustness of various statistical methods, and/or expected variation across different loci. Clearly, more attention needs to be given to examining differences across studies, but the general inference I draw is that the genetic data support an ancient population expansion. What is less clear, however, are the implications to the modern human origins debate.

7.3.4 Estimating the Time of the Expansion

The idea that our species has undergone a dramatic increase in population size is not surprising. Archaeological and historical data show a very rapid increase in the human species over the past 12,000 years after the origin of agriculture. Is *this* the event being detected in genetic analysis, or was there an earlier rapid increase in population size? To attempt reconstruction of our species' demographic history, we need to approximately know *when* the expansion took place.

Alan Rogers and Henry Harpending have developed methods for estimating the timing of an expansion from mitochondrial DNA mismatch distributions.[25] Their investigations showed that the peak of the mismatch distribution (e.g., Fig. 7.6) corresponds to the time of the expansion when the number of differences is expressed in terms of the number of mutational changes. Their basic model postulates a rapid increase in effective size from a small initial population. Mismatch distributions show us the distribution of genetic differences between pairs of individuals, and these distances are proportional to the length of time that connects them. These distances will tend to be longer in large populations than in small ones, because the probability of coalescence is lower in large populations. The peak of the mismatch distribution corresponds to the point in a population's demographic history when the most coalescent events occurred. Rogers and Harpending have shown that there will be few coalescent events after a population expansion, with most occurring immediately before the expansion.

The simplest application of these results is for a model of "sudden expansion." Here, a population of effective size N_0 increases instantaneously to effective size N_1 at t generations in the past. This is obviously an oversimplification, because no population will actually increase "instantaneously" at a single point in time. However, Rogers and Harpending have shown that this simple model accurately captures the timing of the beginning of rapid population growth.[26] There are few differences in simulation studies between the sudden expansion model and more complex models of demographic change, such as exponential growth over time. The simpler model is preferable because there are only three parameters: N_0, N_1, and t.

Unfortunately, we cannot directly estimate these parameters from genetic data. However, we can estimate parameters that are *proportional* to N_0, N_1, and t. We can estimate mtDNA diversity (θ), which is proportional to effective population size. Specifically, the mtDNA diversity expected before the population expansion is

$$\theta_0 = 2\mu N_0$$

where μ is the aggregate mutation rate for the sequence under analysis. Likewise, the expected mtDNA diversity after the sudden expansion is proportional to N_1:

$$\theta_1 = 2\mu N_1$$

Finally, the number of generations in the past when the expansion occurred (t) is proportional to τ

$$\tau = 2\mu t$$

These three parameters provide us with information about population size and time as expressed in units of mutational change. θ_0 and θ_1 measure population size in terms of $1/2\mu$ individuals, and τ measures time in terms of $1/2\mu$ generations. To convert these estimates to actual estimates of size and time, we need to have an estimate of the total mutation rate μ. Given that, we can estimate N_0, N_1, and t as

$$N_0 = \frac{\theta_0}{2\mu}$$

$$N_1 = \frac{\theta_1}{2\mu}$$

$$t = \frac{\tau}{2\mu}$$

Keep in mind, of course, that all estimates of effective population size are for *female* effective size, because of the maternal inheritance of mitochondrial DNA.

Rogers and Harpending developed a method for estimating θ_0, θ_1, and τ from a mismatch distribution. However, this method did not always work, and so they simplified the model further by assuming, given rapid and extensive population growth, that N_1 would often be so large relative to N_0 that, for all practical purposes, it could be assumed to be infinite. If you are not familiar with mathematical modeling, this probably sounds bizarre. Simplifying assumptions are often used in developing mathematical models; these assumptions are often acceptable if you can demonstrate that the simplified model does not distort reality largely. In this case, Rogers found that setting N_1 equal to infinity is a close approximation to the case when N_1 is large. By introducing this assumption, he was able to develop an accurate way of estimating the time since expansion.

Rogers' method starts by computing the observed mean (m) and variance (v) of the mismatch distribution. Given these two easily computed values, preexpansion diversity is estimated as

$$\theta_0 = \sqrt{v - m}$$

and time (in mutational units) since expansion is estimated as

$$\tau = m - \sqrt{v - m}$$

These estimates are then each divided by 2μ to estimate preexpansion population size (N_0) and the number of generations since the expansion (t), respectively.

The estimates of the timing of the Pleistocene population explosion vary across populations as well as by specific estimates of mutation rates that are used to convert τ into the number of generations. In general, however, these estimates tend to fall within the past 100,000 years, with many clustering at about 40,000 years ago. Steve Sherry and colleagues applied this method to data on mitochondrial DNA diversity from 25 different samples from across the world.[27] Significant population expansion was found in 23 of these samples; the other 2 samples (the Hadza and Herero, both in sub-Saharan Africa) are known to have experienced fairly recent bottlenecks that could have resulted in a major reduction in mtDNA diversity, thus "erasing" any previous signal of population expansion. (A recent analysis by Laurent Excoffier and Stefan Schneider also found a lack of expansion among some contemporary hunting and gathering societies, suggestive of recent bottlenecks occurring after the rise and spread of agriculture.[28]) The remaining 23 samples all showed evidence of a rapid expansion within the past 100,000 years, averaging about 40,000 years ago.

Sherry and colleagues also combined populational data into different geographic regions. Their analysis of hypervariable region I (HVRI) data gave estimates of an expansion time of 58,000 years for Africa, 51,000 years for Asia, and 29,000 years for Europe. Another analysis, pooling hypervariable sites I and II, gave estimated expansion times of 64,000 years for Africa, 60,000 years for Asia, and 31,000 years for Europe. Considering the estimates from both local population and regional analyses, these results suggest that the human species expanded rapidly in size roughly 40,000–50,000 years ago, and that sub-Saharan African populations expanded in size somewhat earlier.[29] There is some indication here that the non-African expansion lagged the African expansion.

The mitochondrial DNA data suggest a late Pleistocene population explosion. However, this inference (as with many others) assumes that mtDNA is selectively neutral, so that our analyses are detecting the signature of past demographic history and not a history of past natural selection. Because the issue of neutrality in mitochondrial DNA is still not resolved, any inferences about population expansion would be stronger if we had independent information on the timing of expansion from other genetic traits.

Two studies have estimated expansion time from microsatellite DNA data. As noted earlier, Anna Di Rienzo and her colleagues developed a model relating the variance in DNA repeats to time since population expansion.[30] Applied to their pooled data from three populations, the model estimates that the expansion took place somewhere roughly between 49,000 and 490,000 years ago. The wide range is due to variation in estimates of mutation rate; the younger date corresponds to the higher mutation rate. David Reich and David Goldstein took a different approach in their analysis of the distribution of microsatellite DNA repeats for 60 loci.[31] They used a computer simulation of a rapid expansion and examined the fit of their model to their observed variation in the variance of the number of repeats. This method produced an interval of possible expansion times, ranging from 49,000 to 640,000 years.

Although both of these studies have a wide range of estimates, these ranges do overlap with ranges based on mitochondrial DNA analysis. It is clear that we need more precise estimates of mutation rates and additional loci to narrow the range, but it does appear clear that all of these dates come well before the rise of agriculture (which began about 12,000 years ago). The genetic evidence seems to be telling us that there was an earlier population expansion.

7.3.5 What Happened Before the Expansion?

The genetic evidence reviewed so far in this chapter suggests two findings. First, the long-term average species effective size over the past several hundred thousand years has been small ($N_e \approx 10,000$). Second, there was a large increase in species size sometime during the Late Pleistocene, with most estimates in the range of 40,000–50,000 years (but with large confidence intervals). This expansion has implications for interpreting the small species size. The long-term *average* species size includes periods of time with low size and periods with much larger size. Even though the long-term estimate will lie closer to the smallest size than the arithmetic average (as discussed earlier), this still means that if the rapid expansion model is correct, then the preexpansion species effective size would be less than 10,000. Alan Rogers has suggested that the preexpansion size would be several thousand individuals at most.[32] On the surface, these findings lend support to a replacement model in which modern humans arose as a new species in a relatively small part of Africa and then later expanded in both size and geographic distribution to replace archaic humans outside of Africa. We will return to this suggestion in a bit, but for now, we have to consider what the effective species size was *before* the expansion.

Henry Harpending and his colleagues have considered two basic models of pre-expansion demographic history—either the preexpansion size was large and then plummeted, or the preexpansion size was always (or frequently) small.[33] Harpending refers to these models as the "hourglass model" and the "long-necked bottle" model, respectively, the terms referring to the shape of a graph of population size over time (see Fig. 7.8). In the hourglass model, the species population size is very large for a long time, decreases rapidly in a classic bottleneck, and then increases rapidly afterwards. Under this model, the Pleistocene expansion is a recovery from a bottleneck. In the long-necked bottle model, population size is always small before the Pleistocene expansion.

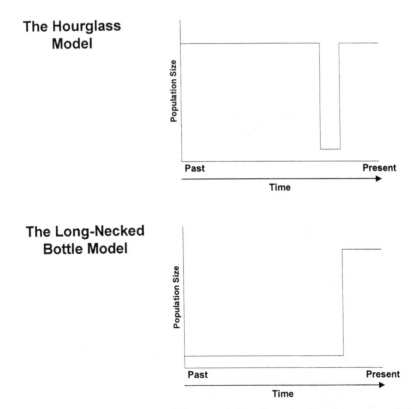

Figure 7.8. *Two models of ancient population growth: The "hourglass" model proposes that a Late Pleistocene population explosion followed a brief bottleneck in population size. The "long-necked bottle model" proposes that population size before the Late Pleistocene population explosion was always small and there was no recent bottleneck. See Harpending et al. (1996, 1998).*

Both models postulate a rapid increase in population size during the Late Pleistocene. What could have driven this demographic change? Several anthropologists have noted that population expansion on the order of 40,000–50,000 years ago corresponds to what some archaeologists have termed the "creative explosion," which is the appearance of many new technologies, such as blade tools and bone tools, and increasing evidence of symbolic expression, such as cave art. If this correspondence is not coincidental, then we may be seeing the demographic consequences of rapid cultural adaptations leading to population growth. As stated by Henry Harpending and colleagues, "The suggestion is that culture rather than biology drove the burst of growth in our ancestors."[34] Of course, this hypothesis assumes that the creative explosion is a real event and not a pattern that we impose on an incomplete archaeological record, seeing abrupt changes instead of smooth transitions. Some archaeologists, such as Geoffrey Clark, argue that there is no evidence for an abrupt disjunction between the Middle and Upper Paleolithic cultures.[35] There is also evidence for bone tools, a cultural trait traditionally associated with the Upper Paleolithic in Europe, to have existed much earlier (90,000 years) in Africa.[36]

The two models differ in terms of what happened before the Late Pleistocene expansion. Under the hourglass model, the expansion is a recovery from a bottleneck that could have been caused by a number of factors, including genetic drift associated with the founding of new populations dispersing out of Africa. Archaeologist Stanley Ambrose offers another fascinating possibility relating to a major geologic event—the eruption of Toba, a super-volcano in Indonesia, at 73,500 years ago.[37] This eruption was the largest volcanic eruption known during the past 450 million years, and it may have led to severe climate change and a "volcanic winter." If so, the impact on human (and other) populations could have been intense, particularly in the northern latitudes. Ambrose proposes that the Late Pleistocene expansion is a recovery from this volcanic-induced bottleneck. Further investigation is needed to address this hypothesis. This model assumes that there was a bottleneck, which fits the hourglass model but not the long-necked bottle model.

Which model of preexpansion demography is correct? We have not sampled enough loci with sufficient time depth to answer this completely, but some evidence does point to the long-necked bottle model. As noted earlier, Rosalind Harding and colleagues' analysis of beta-globins suggests an effective species size of roughly 10,000 back as far as 800,000 years ago. Based on their analysis of *Alu* insertions, Steve Sherry and colleagues estimate an effective species size of 18,000 back as far as one to two million years ago. Based on these studies there seems to be a suggestion that effective species size has been small for a long time, perhaps dating back to the origin and dispersion of *Homo erectus*. Henry Harpending and colleagues have interpreted such results as evidence that our ancestral lineage was always small and a separate species for the past one to two million years. If true, this population history might support the replacement model, in which modern humans are just the most recent of a series of speciation events.[38]

John Hawks and colleagues have suggested an alternative model of ancient demographic history.[39] They argue that the key demographic event was a bottleneck two million years ago with the initial origin of *Homo erectus* (who they label as early

The Exponential Growth Model

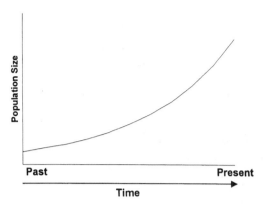

Figure 7.9. *The exponential growth model of ancient population growth proposed by Hawks et al. (2000), in which there was no Late Pleistocene population explosion operating on a small human population.*

examples of the evolutionary species *Homo sapiens*). After this initial origin, species size grew slowly at first and then more quickly over time following a classic exponential growth model (see Fig. 7.9). They suggest that this new evolutionary lineage may have only numbered about 10,000 individuals initially and then grew slowly and gradually to about 500,000 by 800,000 years ago and up to six million by 10,000 years ago. Hawks and colleagues also suggest that archaeological evidence shows a gradual increase in population size and density.

How does the exponential model fit the genetic data? Hawks and colleagues note that the small species size estimated from *Alu* insertion data is consistent with this model. They suggest that the genetic evidence for a later bottleneck (and subsequent expansion) is not that clear. The beta-globin data, for example, show no sign of expansion. They argue that evidence of expansion from mitochondrial DNA and Y chromosome data (which *do* suggest recent expansion) are misinterpreted and that the data most likely are reflecting natural selection. Although microsatellite DNA also shows evidence of expansion, Hawks and colleagues feel that such evidence is inconsistent and contradictory. For example, they note that although the Reich-Goldstein study found evidence of expansion in some African populations, others did not. Another problem noted by Hawks and colleagues is that different genetic systems provide different results; they argue that a recent population bottleneck should be apparent across all genes. However, I think we might also expect a certain amount of variability across systems depending on conditions of initial genetic variation and random effects across loci. A larger number of independent genetic traits will be needed to further evaluate the claims of Hawks and colleagues. For the moment, I prefer to interpret the relatively consistent findings of expansion across mitochondrial and microsatellite DNA as strong evidence of a Pleistocene expansion. Models of modern human origins must consider the possibility of such an expansion.

7.4   EVOLUTIONARY INTERPRETATIONS

Although there continues to be some debate over the genetic evidence regarding ancient demography, I suggest that at present there are several basic findings that must be considered in any evaluation of modern human origin models. First, the long-term average effective species population size of our ancestors over the past several hundred thousands years is roughly 10,000, much smaller than any ecologically based estimates, which range from a few hundred thousand to more than a million for this time period. Second, there is genetic evidence that our species grew rapidly in size at some point within the Late Pleistocene. Third, there is some (although less convincing) evidence that suggests that the expansion began somewhat earlier in Africa than elsewhere in the Old World. Each of these findings can be criticized, particularly in terms of neutrality assumptions. However, to evaluate completely different origin models, we need to examine the possibility that these findings are accurate reflections of reality and then determine appropriate interpretations as well as directions for future research.

7.4.1   Evidence of Speciation and Replacement?

Henry Harpending and others have argued that the low estimated species effective size of roughly 10,000 adults is incompatible with a multiregional interpretation, and a recent African replacement is more likely to be correct. If adults make up roughly half of a hunting and gathering population, then an effective population size of 10,000 is roughly equivalent to a census size of 20,000. This estimate is considerably smaller than ecologically based estimates that range from several hundred thousand to over a million.

Why are these two estimates so different? One argument is that they measure different things. The ecological estimates are based on the total number of hominids that were likely to have lived throughout the Old World given an expected range of population densities. This estimate would include *all* hominids, regardless of whether they were ancestors of modern humans or not. If, for example, the European Neandertals were a separate species that made no genetic contribution to later human evolution, they would still be counted in any ecological estimate that relied solely on location. After all, we know that early humans such as Neandertals *did* live in Europe; the argument is whether they made any genetic contribution to living humans. The genetic estimates, on the other hand, only include those individuals that actually are our ancestors. Extinct lineages would not be counted. Thus one explanation for the difference between ecological and genetic estimates of species size is that our genetic ancestors account for only a fraction of the hominids that lived in the Old World over the past million years or so. The other hominids became extinct.

Why should small species size reject a multiregional model? A key difference between replacement and multiregional models is the suggested geographic spread of our ancestors 200,000 years ago. According to the replacement model, our ancestors were confined to a single region—sub-Saharan Africa—at this time. Hominids living elsewhere were ultimately replaced. Multiregionalism, on the other hand, predicts that our ancestors lived in more than one region in the Old World. If we assume that a species effective size of 10,000 corresponds roughly to a total census size of

20,000 ancestors on average over the past few hundred thousand years, then we can make some estimates of geographic distribution based on estimates of likely population density.

As noted earlier, archaeologist Fekri Hassan estimates that population density during the Middle Paleolithic would be roughly 0.03 persons per square kilometer. Applying this density to a census size of 20,000 people gives a total occupied land area of roughly 0.7 million square kilometers, roughly equal to the size of the state of Texas. This is a very small area, and it suggests that our ancestors were limited to a small geographic range, which is compatible with the idea that *Homo sapiens* originated as a new species in a small, localized population and then spread out across the Old World replacing preexisting archaic humans.

On the surface, small species effective size seems incompatible with a multiregional interpretation, which requires that our ancestors occupied more than one geographic region. Hassan, for example, argues that roughly 40 million square kilometers of the Old World was occupied during the Middle Paleolithic. If *all* of these populations were ancestors of modern humans and if the estimate of 20,000 people is even approximately correct, then they would have been incredibly spread out! Assume, for example, small local bands of 25 people. If there were 20,000 people, this means there would be 800 bands spread out over 40 million square kilometers, or one small band every 50,000 square kilometers. If these bands were spaced out in a grid, this means that there would be roughly 230 kilometers separating each band. According to Harpending and colleagues, there were too few people to be spread out across the entire Old World and still allow frequent enough contact to ensure gene flow within a single species. If we grant all of the above assumptions, then the evidence for small effective species size essentially rejects a multiregional model and supports a replacement model.

Of course, these numbers are all rough approximations useful only for general insight. Still, the evidence for small effective species size suggests rejection of a multiregional model even under less limiting conditions. If, for example, the census size was twice as large as our estimates (40,000) and population density was as low as 0.01 persons per square kilometer (as has been found in some living hunting-gathering societies), then we would still only have a total geographic area of four million square kilometers, which is one-tenth of Hassan's estimate.

However, it is also clear that changes in one of more of our estimates can result in a wide range of total geographic area. We must take into account that fact that each of the parameters we have looked at has a range of variation. Alan Templeton has noted that the ratio of breeding size to census size is frequently as low as 0.2 in other large-bodied mammalian species. If we use this estimate, rather than the estimate of 0.5 used so far, an effective population size of 10,000 implies a census size of $10,000/0.2 = 50,000$. Using this estimate increases the geographic range to roughly 1.7 million square kilometers with a population density of 0.03 people per square kilometer and 5.0 million square kilometers with a population density of 0.01 people per square kilometer. Perhaps the largest source of variation is in the genetic estimates of effective species size. Such estimates have very large confidence intervals. Templeton suggests that estimates of effective species size based on mitochondrial DNA are compatible with a census size of between 200,000 and 500,000 people. Looking at estimates from nuclear DNA, John Hawks and colleagues note that the upper confidence intervals for species effective size range as high as

18,000–160,000, averaging about 70,000 for seven different genetic systems. This average implies a census size of 140,000–350,000 people, depending on the ratio of breeding to census size used in computation. The point made by these authors is that although genetic and ecological estimates of census size appear at first glance to be quite different, this difference becomes less and less as we factor in sources of variability.

Although it is true that individual genetic estimates of species effective size *do* have large confidence intervals, it is interesting that most systems agree on a figure close to 10,000, which suggests that the "true" figure is close to this average than to the upper limit. We also have to deal with the fact that all of these estimates (size, geographic area, etc.) are based on a single, constant value averaged over the past few hundred thousand years. If we accept the genetic evidence that there was a Pleistocene population explosion, then the small average long-term species effective size reflects a period during which the *initial* size was much less, perhaps only a few thousand individuals. Once again, many would argue that all of this evidence collectively points to a very small number of ancestors roughly 200,000 years ago, in line with a replacement model. Even if we increase this small number by an order of magnitude, we still don't have enough people to have occupied the entire Old World.

If for the moment we accept that the genetic evidence implies both a small species census size and a limited geographic distribution, then what inferences can we make? These data strongly support a replacement model. In this view, modern humans arose as a separate new species, *Homo sapiens*, in part of Africa roughly 200,000 years ago. By 100,000 years ago, some populations had dispersed out of Africa into the remainder of the Old World. Between 100,000 and 50,000 years ago, these regional populations began expanding rapidly in size, first in Africa and then elsewhere. During this process, preexisting archaic human populations outside of Africa were replaced, presumably because of some biological and/or cultural advantage in the new species. In other words, the data on small species size, Pleistocene population growth, and higher diversity in Africa (which suggests earlier expansion) fit Harpending's "weak Garden of Eden" model.

According to this view, the *key* genetic evidence supporting replacement is the low species effective size. Many other genetic arguments, such as levels of African diversity and the age of "mitochondrial Eve," have too many alternative explanations that make them also compatible with a multiregional explanation. The small species size, and the implied low geographic range of our early ancestors, provides the best single argument in favor of replacement and rejecting a multiregional explanation. The fact that other genetic data are *compatible* with replacement is reassuring but they are not sufficient by themselves to settle the matter.

When I first began research into the genetics of modern human origins, the above paragraph essentially summarized my views. The developing evidence for small species size convinced me that earlier genetic claims for replacement were correct, but for the wrong reasons. It was not the age of "Eve" or the diversity of sub-Saharan Africans that settled the issue, but the small species size. However, it also occurred to me (and many others) that the situation might not have been that simple. At that point, the primary evidence for small species size came from mitochondrial DNA, which meant that we could make inferences about average long-term species size only back as far as the date of coalescence, which averaged about 200,000 years. What was species size like before that time? Was it possible that our ancestors were much more numerous and then, for whatever reason, suffered a major bottleneck bringing their numbers down? Perhaps the expansion from a small size was not the signature of the birth of a new species, but rather a population bottleneck within a single species. This idea is the "hourglass" model posed by Harpending and colleagues. The alternative, the "long-necked bottle" model suggests that species size was very small for a long time, perhaps representing multiple speciation events over the past two million years. This idea fits the suggestion by some paleoanthropologists that modern *Homo sapiens* is just the most recent in a series of Pleistocene speciation events, including *Homo ergaster*, *Homo antecessor*, and *Homo heidelbergensis*. We still need more traits with older coalescence times, but the evidence to date from beta-globins, *Alu* insertions, and other traits suggests that the long-necked bottle model might be more appropriate. Small species size is then a function of phylogenetic history, with multiple speciation events, rather than a demographic bottleneck within a single evolutionary lineage.

As I think back about these developments, it seems to me that I should have become even more firmly convinced of speciation and replacement. At the same time, however, I began to wonder if all of our assumptions were necessarily correct. In particular, what is the relationship between species effective size, breeding size, and census size? As I pondered this question, I became aware of another possibility, having to do with the influence of extinction and recolonization of local populations.

7.4.2 Extinction and Recolonization of Local Populations?

The famous population geneticist Sewall Wright addressed the question of the relationship of effective population size and census population size in a 1940 paper in *American Naturalist*. In his discussion of the evolutionary importance of fluctuations in population size, Wright noted "An important case arises where local populations are liable to frequent extinction, with restoration from the progeny of a few stray immigrants. In such region the line of continuity of large populations may have passed repeatedly through extremely small numbers even though the species has at all times included countless millions of individuals in its range as a whole."[40] Consequently, effective size can be much less than census size.

This process has since been studied in more detail and forms a major focus of the field of metapopulation biology, which concerns itself with the evolutionary effects of a population subdivided into small local populations that frequently become extinct.[41] The underlying model is one of extinction and recolonization of these local populations. A species occupies patches of the environment. When a local population becomes extinct, the empty patch is recolonized by members from a nearby, related population. This process can maintain a large census size, but because of genetic drift introduced by colonization of small numbers from related populations, the overall *effective* size is often quite small. The dynamics of extinction and recolonization are crucial to conservation biologists, who must deal with reduced genetic variability associated with small effective species size.

The process of extinction and recolonization is easily illustrated with a simple simulation. Figure 7.10 shows five populations (labeled A to E), each occupying a given area. I used a random number generator to simulate the extinction of each

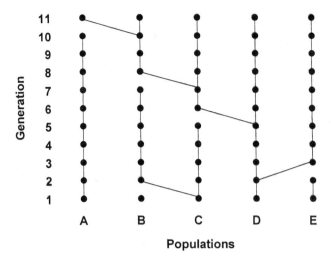

Populations

Figure 7.10. Simulation of the extinction and recolonization of local populations. Each generation the five populations (A–E) were each allowed to either survive or become extinct (with probability = 0.1 per population per generation). If the population became extinct, its habitat was recolonized from an adjacent population (chosen by random number). Vertical lines indicate survival, and diagonal lines indicate recolonization from an adjacent population. After 11 generations, all 5 populations trace their ancestry back to a single population (D). All other lineages have become extinct. In this model, all other populations have been replaced.

population in each generation, setting the probability of extinction equal to 0.1. For each population in each generation, I picked a random number that ranges from 0 to 1. If the number was greater than 0.1, the population survived to the next generation. If the number was less than 0.1, then the population became extinct, and its area was colonized by one of the adjacent populations (also chosen using a random number, unless there was only one adjacent population). In the simulation in Fig. 7.10, population B became extinct in the first generation and its area was colonized by migrants from population C in the second generation. Population E became extinct in the second generation and was colonized by migrants from population D in the third generation. No populations became extinct in the third or fourth generations, and then population C became extinct in the fifth generation and was replaced by colonists from population D in the sixth generation. The process is shown for 11 generations, by which time all 5 populations are directly related back to population D.

In this example, the process of extinction and recolonization of local populations results in the total replacement of all five geographic areas from population D. Assuming no migration between the five populations, it is clear that all of the genes in this species in generation 11 trace back to population D. Naoyuki Takahata has argued that this type of process was responsible for the origin of modern humans. He states, "local populations in Eurasia might have undergone more frequent extinction than did those in Africa, because of the then-adverse environment. Africans could then serve directly or indirectly as founding populations in Eurasia."[42] This gradual extinction process is an alternative to a rapid replacement across the Old World.

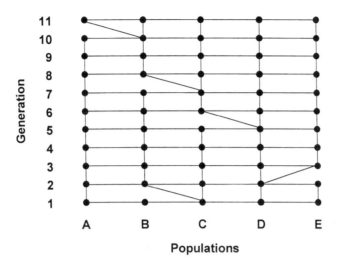

Populations

Figure 7.11. *Simulation of the extinction and recolonization of local populations under gene flow. This is the same result as in Fig. 7.10, but gene flow between populations is allowed each generation (indicated by horizontal lines). After 11 generations, all 5 populations can be traced back to population D, but because of gene flow each of these 5 populations also has ancestry from the other 4 populations. Under this type of model, most ancestry would come from population D but not all of it. Additionally, this process will result in an effective species size much lower than the species census size.*

A major problem with this model, however, is that it assumes that there is no migration between local populations, so that ultimately every population is uniquely descended from a single population. What happens when there is migration between local populations in each generation, as shown in Fig. 7.11? In this model, genes are shared between local populations as well as through the process of extinction and recolonization. Although all populations in Fig. 7.11 still trace their ancestry back to population D, the actual genetic composition of the populations has been mixed each generation through the process of migration. Thus each population has multiple ancestries, in line with a multiregional model. The question here is how extinction and migration balance one another and the expectations of effective species size under a multiregional model.

A number of theoretical treatments of local extinction and recolonization have been produced.[43] One of the most general models has been developed by Michael Whitlock and N. Barton. Their model takes into account the number of local populations, the breeding size of each population, the probability of extinction, the migration rate between populations, the number of initial colonists repopulating an area, and the genetic relationship between these colonists. All of these factors affect both effective population size and the amount of genetic differentiation between groups (F_{ST}). Whitlock and Barton show how species effective population size can be expressed as a function of the breeding population size of local populations, the number of local populations, the level of F_{ST} among populations, and the per-generation probability of a local population becoming extinct and being replaced with migrants from another local population.[44]

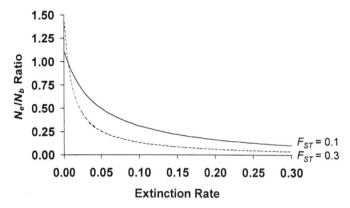

Figure 7.12. *The impact of extinction on the relationship between effective population size and census population size using the Whitlock–Barton model with local breeding population size of 100 and an F_{ST} of 0.1 (solid line) and an F_{ST} of 0.3 (dashed line). At zero level of extinction, effective size is actually larger than census size. For even small rates of extinction and recolonization, the ratio drops very quickly and effective size is much lower than census size.*

Figure 7.12 shows the effect of different values of extinction rate and F_{ST} on the ratio of effective population size to breeding population size. This example uses a species made up of local breeding population sizes of 100 for two values of F_{ST} (0.1 and 0.3). For very low (or zero) levels of extinction, the ratio of effective to breeding size is actually greater than 1.0, indicating that, in the absence of extinction, genetic differentiation inflates effective population size.[45] However, note that this situation rapidly changes for even slightly larger extinction rates. In this example, an extinction rate of 0.05 and $F_{ST} = 0.1$ results in an effective to breeding size ratio of 0.5. In other words, an extinction/recolonization rate of 0.05 reduces the effective size to one-half the breeding size under relatively low differentiation of populations ($F_{ST} = 0.1$). In this case, the actual breeding size in each population is 100, but the effect of extinction and recolonization is to make the population genetically equivalent to 50 adults. For a higher level of differentiation ($F_{ST} = 0.3$), the ratio drops to 0.26 for the same extinction/recolonization rate. The larger the extinction rate and/or F_{ST}, the greater this effect. An extinction rate of 0.2 and $F_{ST} = 0.3$ results in a ratio of 0.06. Each local population would be genetically equivalent to six adults.

Keep in mind that these examples refer to the ratio of effective size to breeding size. Because the number of reproductive adults in a population is less than the total census size, the ratio of effective size to *census* size would be even smaller. If, for example, the number of reproductive adults made up one-half the total census population, then the ratio of effective size to census size would be one-half the above ratios.

What can we say about the likely effect of extinction and recolonization of local populations in recent human evolution? Unfortunately, we cannot make direct quantitative predictions about the exact ratio of effective size to census size because we do not have exact estimates of either F_{ST} or extinction rates in human prehistory. Nonetheless, we can make a qualitative case for conditions favorable for a low ratio. Given what we know about the fossil and archaeological record, we need to consider the possibility that F_{ST} was higher in the past than today, particularly when we consider the likelihood that populations were more fragmented during past climatic shifts (e.g., glaciations). We also know from Alan Templeton's work discussed in Chapter 6 that F_{ST} for *living* humans is very low relative to other large-bodied mammals. If F_{ST} for our ancestors were more similar to that of other living mammal species, then we would expect a lower ratio of effective to breeding size.

We know little about the per-generation probability of the extinction of a local human population. Extinction of local populations can be caused by random demographic fluctuations (e.g., fewer female births by chance) as well as ecological shifts (e.g., climatic shifts) and changes in mortality patterns (e.g., epidemics, aggression), among others. It is most likely that extinction probabilities have changed over both time and space, although we need only an average level for the Whitlock–Barton model. The problem is that we have little information on the exact levels of local population extinction from our past. Joseph Soltis and colleagues provide some information on human extinctions in small Papua New Guinean populations.[46] Extinction rates per generation range from 0.016 to 0.313. However, these populations are horticultural groups with frequent intergroup warfare, and it is not clear to what extent they can be used to model earlier hunting and gathering populations. Studies of the age-at-death distribution of fossil remains suggest that prehistoric human populations may have been more prone to extinction than among living humans. Erik Trinkaus investigated the demographic profiles of Neandertals and found that the age-at-death distribution suggests "a profile that is at or beyond the limits of recent human demographic profiles," which implies "a population under severe demographic stress, primarily through the exceptionally high prime-age adult mortality rates and a consequent dearth of older individuals."[47] Although such evidence supports the idea of higher local population extinction rates in the past, we still do not know the exact magnitude. We will need further ethnographic and archaeological data to narrow this range. For the moment, we can say that conditions suitable for a small ratio of effective to census size, such as fragmentation of populations and adverse conditions for survival, were present during the Pleistocene. It seems likely that humans, as with most other mammalian species, are characterized by a low effective size despite a large census size. If so, then we cannot draw any conclusions about the *actual* number of people in the past from genetic data and therefore cannot support a model of African replacement over a multiregional alternative.

The possibility of extinction and recolonization of local populations might also play a role in the observed Late Pleistocene population expansion. My colleague Elise Eller has suggested that this population expansion may be caused in part to a decline in local population extinction.[48] If these rates declined because of improved cultural adaptations, for example, then the species effective size would increase without any necessary increase in species census size. This idea views local extinction and recolonization as having a damping effect on species effective size. When this effect is reduced, the species effective size increases closer to the breeding population size. The same effect could also be caused by a reduction in F_{ST}, perhaps relating to increased migration among populations, which in turn could reflect changing environmental factors and/or cultural adaptations. Although there is no doubt among many archaeologists that the actual census size *did* increase, it is also possible that the increase in species effective size was in part related to changes in extinction and population differentiation. Under this model, the Late Pleistocene population explosion in *effective* population size reflects 1. An increase in *census* size, 2. A reduction in the rate of local extinction and recolonization, *and* 3. A reduction in the degree of population differentiation (lowering F_{ST}). This model needs to be investigated more closely in future genetic and archaeological analyses.

7.5 SUMMARY

Although it seems strange at first, many of the arguments over genetic inferences about modern human origins are demographic—how many ancestors did we have? It is clear from the archaeological and fossil records that humans, defined in the broadest sense, have been spread out over parts of Africa, Asia, and Europe during much of the past two million years. The key question is, Which ones were our biological ancestors? If we could trace each of our family trees back through the past 100,000 to 200,000 years, where would we find our ancestors? Did they *all* live in Africa? If so, then other humans, such as the Neandertals, are not our ancestors and died out for one reason or another. On the other hand, did our ancestors live across the entire Old World, or perhaps only in some parts?

The answer to these questions is essentially demographic—how many ancestors? We can't simply estimate how many early humans existed, because we would then have to assume that they were all ancestral to living humans, the very question we are trying to answer. Our question has to do with the *genetic* number of ancestors, a quantity that can be estimated from genetic data using a variety of different methods. The answer to date is fairly consistent, equating to a population of roughly 10,000 adults. If we take this figure as a *direct* estimate of the breeding size of our species, then we are faced with the simple conclusion that there were too few people to have been spread out over a large geographic area. Given what we know about hunting-gathering population density, these few adults would have lived within a localized part of Africa and could not have been spread out over the entire Old World. Given this logic, the genetic data can arguably be used to support African replacement and rule out a multiregional interpretation.

A potential problem with this logic is equating the effective species size directly to the breeding species size. Was there actually only an average of a few thousand adults, or do patterns of genetic variation merely give this impression? Effective population size can be *much* less than breeding (or census) size if there was frequent extinction and recolonization of local populations. It seems *possible* that our ancestors numbered in the hundreds of thousands but had genetic variation reduced through the extinction/recolonization process, giving the impression that there were far fewer ancestors. We don't have enough direct information on extinction rates, or other parameters, to determine whether this actually *did* happen, but it is clearly a possibility. In sum, our species' low effective size (~10,000) does not necessarily rule in favor of a replacement model.

Genetic data also suggest a rapid expansion in effective population size during the Late Pleistocene. This rapid growth could be a reflection of speciation and subsequent dispersal and growth, as predicted from a replacement model. It might also reflect population growth within a single evolutionary lineage, as predicted by a multiregional model. A related question is, What drove this growth? Are we seeing the genetic signature of rapid cultural changes, leading to higher population densities and size, or a reflection of new genetic adaptations producing the same result? In both cases, we could argue that such changes took place within a new species or within a much larger geographically widespread lineage. Resolution of these questions relates back to the issue of the actual numbers of ancestors. As with the genetic inferences discussed in earlier chapters, the data are indeterminate on their own. We need to consider low effective species size along with the other genetic data, as well as the fossil and archaeological evidence.

Chapter 8

Neandertal DNA

All of the genetic evidence discussed so far refers to inferences made from patterns of genetic variation obtained from *living* humans. This type of inference fits the framework described in Chapter 1, that our species' *current* genetic variation is a reflection of the past. We can easily measure and describe genetic variation among living humans and then turn to the question of which model provides a better explanation of how we got this way. This approach seems reasonable given that living people are the source of our data, and, lacking a time machine, how could we obtain information on genetic variation in the *past*?

At first, the idea of looking at ancient genes may not seem difficult. After all, you have probably seen the movie (and/or read the book) *Jurassic Park*,[1] in which scientists extract ancient dinosaur DNA from insects trapped in amber. The movie contains a scene in which this process is described in a short cartoon. A prehistoric insect bites a dinosaur and drinks some blood. The insect then lands on a tree and is stuck in tree sap, which eventually hardens and fossilizes into amber. Millions of years later, the *Jurassic Park* scientists extract the dinosaur blood from the insect and reconstruct dinosaur DNA sequences. The DNA sequences are completed using frog DNA to fill in the missing pieces (although bird DNA would have been a better choice, given that dinosaurs are more closely related to living birds than to living amphibians—still, let's not forget that this is, after all, a movie). The end result—replicated dinosaurs that predictably escape and go on a rampage.

8.1 STUDYING ANCIENT DNA

Although it is quite entertaining, *Jurassic Park* is science fiction. However, much of the storyline is based on science fact. It *is* possible to extract and sequence some pieces of ancient DNA (aDNA), although not enough for more than a minute fraction of an organism's total genetic code.[2] Ancient DNA analysis is made possible by a laboratory method known as the polymerase chain reaction (PCR). In simplified terms, the polymerase chain reaction starts by separating the DNA in a sample into two strands, which are then used as templates for the synthesis of two new strands, thus duplicating the amount of original DNA. The new double-stranded DNA sample is then split into separate strands and the process is repeated, yielding even more DNA. The entire process is repeated in this cyclical fashion, doubling the amount of DNA each time through the cycle. PCR allows very small amounts of DNA to be amplified millions of times. The amplification is so powerful that caution must be exercised to avoid magnifying the DNA of anyone working in the lab from dust in the lab or other contaminants.

PCR is routinely used in all genetic analysis, including DNA from living and ancient organisms. The study of ancient DNA is further complicated by the degradation of DNA over time. Mitochondrial DNA generally holds up better than nuclear DNA, because there are more copies of mitochondrial DNA in cells than nuclear DNA. *If* preservation conditions are favorable, and *if* there is no contamination, it is sometimes possible to extract some mitochondrial DNA from fossils. Most successful ancient human DNA research has been on relatively recent fossils, generally within the past few thousand years. The initial development of ancient DNA research led to an interesting question—would we be ever able to directly sample the genetic code of much more ancient humans, such as the Neandertals?

8.2 DISCOVERY OF NEANDERTAL DNA

As of the summer of 2000, mitochondrial DNA has been extracted from two Neandertal fossils, one from Western Europe, and the other from the northern Caucasus.[3] These extractions represent astonishing technical achievements and show significant differences between the DNA of Neandertals and living humans. Although some of the scientific community, and much of the media, have taken this as the final answer in the debate over the fate of the Neandertals (they became extinct), others are not as certain. There is little debate over the basic findings but much more regarding the evolutionary significance of these findings. Were Neandertals a separate species, long separated from the lineage that gave rise to modern humans? Were Neandertals a relatively isolated subset of humanity (perhaps a different subspecies) that ultimately mixed with other human populations? Does our species today have any Neandertal genes? If so, is there a small amount or is it more significant? What do these findings tell us about the broader question of modern human origins?

8.2.1 The First Neandertal DNA Sequence (Feldhofer)

In the summer of 1997, the time barrier was broken with the first successful extraction of ancient mitochondrial DNA from a human who lived more than 30,000 years ago, as reported in the July 1997 issue of the journal *Cell*.[4] Mathias Krings and colleagues extracted mitochondrial DNA from the arm bone of the first discovered Neandertal specimen, from Feldhofer Cave in the Neander Valley, Germany, who lived roughly 35,000–70,000 years ago.[5] Krings was able to reconstruct a 378-base pair sequence. Keep in mind that this is a very small fraction of the total mitochondrial DNA, which in turn is a very small fraction of the total DNA of the Neandertal. Don't expect this person to be cloned (or the construction of "Pleistocene Park")! The available information is useful, however, in the modern human origins debate, particularly the question of how much (if any) the Neandertals of Europe contributed to the gene pool of modern humans.

Krings and colleagues compared the Neandertal DNA sequence to 994 distinct DNA sequences from living humans across the world. This is the pairwise comparison method discussed in earlier chapters—take any living human DNA sequence and compare it to the Neandertal DNA sequence, counting the number of nucleotide differences, and then repeat this procedure for each of the sequences from 994 living humans. They found an average of 27 DNA differences between living humans and the Neandertal specimen. The individual differences ranged from a low of 22 DNA differences to a maximum of 36 DNA differences. For comparison, Krings and colleagues looked at variation among living humans only, that is, comparing each of the sequences from living humans to the other 993 sequences from living humans. This comparison shows an average of 8 DNA differences, ranging from a low of 1 difference to a maximum of 24 differences.

We see that the average sequence difference between living humans and the Neandertal specimen (= 27) is over three times the average difference between living humans (= 8). There is some overlap, in that there are a few comparisons between living humans that show a greater difference (= 23 or 24) than found between the smallest difference (= 22) between the Neandertal and living humans. However, this overlap is minimal—only 0.002 % of the comparisons between living humans were larger than the smallest Neandertal-living human difference. Although caution was expressed about making wide inferences based on a *single* Neandertal specimen, the results did suggest that Neandertals are rather genetically dissimilar from living humans. The evolutionary meaning of this difference is less clear. Does it mean that Neandertals were different from us because they belonged to a separate species, or because the Neandertals lived tens of thousands of years ago?

In 1999, Krings and colleagues succeeded in extracting another mitochondrial DNA sequence from the Feldhofer specimen.[6] In the first study, the DNA sequence was from what is known as the first hypervariable region (HVR1) of mitochondrial DNA; in the 1999 study Krings reported results based on an additional 340 base pairs from the second hypervariable region (HVRII). In general, the results were very similar to those reported in the 1997 study. For HVRII, the Neandertal specimen is different from living humans at an average of 35 DNA sequence positions (with a range of 29–43 differences). This difference is much larger than found between all pairs of living humans, in which the average DNA sequence difference is 11 (with a range of 1–35 DNA differences). Again, there is some overlap, but not very much. The 1999 study confirmed that the Neandertal specimen was genetically different.

8.2.2 The Second Neandertal DNA Sequence (Mezmaiskaya)

Although debate has continued over the evolutionary implications of the recovery of the Feldhofer mtDNA sequence, mtDNA was extracted from a *second* Neandertal specimen. Igor Ovchinnikov and colleagues reported this discovery on March 30, 2000 in the British science journal *Nature*.[7] This second specimen was that of a

Nucleotide Position

Position	Reference	Mezmaiskaya	Feldhofer
16078	A	A	G
16086	T	C	T
16093	T	T	C
16107	C	C	T
16108	C	C	T
16111	C	C	T
16112	C	C	T
16129	G	A	A
16139	A	T	T
16148	C	T	T
16154	T	T	C
16156	G	A	G
16169	C	T	T
16182	A	C	A
16183	A	C	C
16189	T	C	C
16209	T	C	C
16223	C	T	T
16230	A	G	G
16234	C	T	T
16244	G	A	A
16256	C	A	A
16258	A	A	G
16262	C	T	T
16263.1	-	A	A
16268	C	T	T
16279	A	G	G
16311	T	C	C
16320	C	T	T
16344	C	T	C
16362	T	C	C

Figure 8.1. Comparison of 345-bp mitochondrial DNA hypervariable region I (HVRI) sequences from the Anderson reference sequence for living humans and the two Neandertal sequences obtained to date (Feldhofer and Mezmaiskaya). This figure shows only those nucleotide positions where one or both Neandertals differed from the Anderson sequence (Ovchinnikov et al. 2000). Note that position 16263.1 represents an insertion (nucleotide A) that is present in Neandertals but not in living humans.

Neandertal infant from Mezmaiskaya Cave in the northern Caucasus, representing the eastern range of Neandertals. The fossil was dated to 29,000 years ago, making it one of the more recent Neandertal fossils discovered to date.

The first step in their analysis was to determine that there was molecular preservation. Their preliminary analysis found high enough levels of collagen, a protein in bones and connective tissues, to show good preservation of molecules. PCR was used on samples from two rib fragments to amplify the mitochondrial DNA sufficiently to allow sequencing. As with the analysis of the Feldhofer specimen, the sequencing was repeated in two independent laboratories. The result was a 345-base pair sequence from the first hypervariable region (HVRI) of mitochondrial DNA.

Several comparisons were made with other mtDNA sequences (see Fig. 8.1). Compared with the standard human reference sample (known as the Anderson reference), the Mezmaiskaya DNA sequence is different at 23 nucleotide sites, a large difference.[8] There are 12 differences when compared to the other (Feldhofer) Neandertal sequence (which itself differs from the Anderson reference at 27 sites). As with the Feldhofer sequence, the DNA from the Mezmaiskaya specimen is noticeably different from modern humans. What does this difference mean?

8.3 EVOLUTIONARY INTERPRETATIONS

We are back to a basic, and long debated, question—were the Neandertals a separate species or were they part of our species' ancestry? The comparisons of the two Neandertal DNA sequences with living humans shows that there is a rather large difference that might be a reflection of separate species. The Neandertal specimen falls outside most of the DNA variation we see in our own species today. The genetic evidence certainly suggests the same conclusion that many see in the fossil record—the Neandertals were different. The question is, how different?

There are two different interpretations of the Neandertal DNA evidence that need to be considered in relationship to the modern human origins debate. First, the Neandertals were a separate species that were completely replaced by modern humans moving into Europe. This interpretation does not rule out cultural contact between modern humans and Neandertals, or even mating between the two groups, as long as they had become genetically different enough to be a separate species, incapable of producing fertile offspring. Note that if this interpretation is correct, it does not necessarily imply that replacement occurred elsewhere in the Old World. We can consider, for example, the possibility that the Neandertals were replaced but that there was genetic continuity in Asia. Because ancient DNA is so far available only for Neandertals, the discussion here will focus on the more specific issue of Neandertal replacement, returning to the global questions of modern human origins in Chapter 9.

The second interpretation is that Neandertals were not a separate species but part of a single species spread out across the Old World, as expected under a multiregional model. If the Neandertals and other archaic humans all belonged to the same species, then there are several subsequent questions focusing on the degree of genetic interrelationships with other populations. Was there regular gene flow connecting Neandertals with other regional archaic populations, or was there admixture between Neandertals and incoming migrants from elsewhere? The questions here focus on the nature of gene flow, but all consider that Neandertals were not reproductively isolated from other archaic human populations and contributed *some* genes to living humans. If so, how much?

Any resolution of these questions must deal with observed and expected relationships across time and space. We need to have data on DNA sequences from 1. Living human populations, 2. Neandertals, and 3. Fossils generally accepted as "anatomically modern." Given these data, there are several relevant questions:

- What is the genetic and evolutionary relationship between Neandertal populations?
- What is the genetic and evolutionary relationship between Neandertals and living human populations?
- What is the genetic and evolutionary relationship between Neandertals and contemporaneous "modern" humans?
- What are the genetic and evolutionary relationships between ancient "modern" humans and living humans?

In terms of mitochondrial DNA sequences, we have data on thousands of living humans across the world, data from two Neandertal specimens, and no data (yet) on ancient anatomically modern humans (such as Cro-Magnon). This means that direct testing via mitochondrial DNA sequences limits us to the first two questions—comparing Neandertals to each other and comparing Neandertals to living humans.

8.3.1 Variation within Neandertals

Now that there are two Neandertal DNA sequences, we can *begin* to address questions of genetic variation within the Neandertals. Do the two Neandertal sequences represent the same evolutionary group? Ovchinnikov and colleagues found 12 nucleotide differences between the Feldhofer and Mezmaiskaya sequences. To put this number into perspective, they also examined levels of mtDNA variation *within* geographic regions of living humans for the same 345-base pair sequence. They found that the average pairwise difference between living Europeans and living Asians was 5.3 and 6.3 respectively; in both cases, the difference between the two Neandertal sequences (= 12) is statistically significant. Fewer than 1% of the comparisons between pairs of living Europeans or between pairs of living Asians differ as much as this. On the other hand, comparison with living humans from Africa shows a larger average pairwise difference of 8.4, and the comparison with the Neandertal difference of 12 is not statistically significant. Among living Africans, 37% of the pairwise differences were at least this different.

All told, these analyses show that the difference between the two Neandertal sequences is greater than found among some groups of modern humans (Europe, Asia), but not all (Africa). It seems reasonable to conclude that the two Neandertal sequences represent individuals within the same evolutionary lineage. The fact that there are 12 differences is not surprising given that the two samples do not sample the Neandertals at the same point in time; while the Mezmaiskaya specimen is dated to 29,000 years ago, the dating of the Feldhofer specimen is less certain, with estimates ranging from 35,000 to 70,000 years ago. The two Neandertal specimens thus lived between 6,000 and 41,000 years apart, allowing for the possibility of DNA change over time, increasing our estimate of variation *within* Neandertals. Still, the magnitude of this difference still fits comfortably within the range documented for living African populations, and we can state with fair certainty that the two specimens represent the same evolutionary line. Of course, this all assumes that the Neandertal morphology and dating of the Mezmaiskaya specimen is confirmed. The close similarity of the two Neandertal DNA sequences also provides further support for their accuracy—the first one was not a fluke.

8.3.2 Were Neandertals a Separate Species?

Even though we have only two Neandertal sequences, the data to date show them to be quite different from living humans. To put these numbers into perspective, we need to consider how much variation we would expect *within* and *between* living species. Two sources of comparative data are available for these DNA sequences—living humans and living apes. Comparisons with living humans are done in two ways. One method of comparison is to use a single living human mtDNA sequence known as the Anderson human reference standard. The other method involves pairwise comparisons with sequences from a number of living humans. The results in both cases are almost identical. Comparisons with apes have used DNA sequences from our closest living primate relatives—chimpanzees and bonobos.

As noted above, the initial study of hypervariable region I (HVRI) in the Feldhofer specimen showed 27 differences out of 379 base pairs (7.1%) relative to the Anderson human reference. The number of differences obtained by pairwise comparison with 994 sequences from living humans gave virtually the same result (27.2 differences, with a range from 22 to 36). How many differences would we expect to see if Neandertals and living humans were the same species? To answer this, Krings and colleagues looked at the average pairwise difference between all 994 sequences of living humans. They found an average of 8.0 sequence differences, with a range of 1–24. The difference between Feldhofer and living humans is over three times the magnitude found within living humans (27.2/8.0 =

3.4). Given the low degree of overlap in the distributions, the Feldhofer–living human difference seems well outside of the range of normal variation expected within a species. One conclusion is that this large difference means that Neandertals were a separate species.

Krings and colleagues also looked at mitochondrial DNA differences between 986 sequences from living humans and 16 sequences from living chimpanzees, based on the 333 base pairs in common between humans and chimps. Using only these positions, they found 25.6 differences between Feldhofer and living humans (range 20–34). For comparison, they found an average of 8.0 differences among pairs of humans (range of 1–24) and an average of 55.0 differences between humans and chimps (range of 46–67). The Neandertal–human difference is again roughly three times (25.6/8.0 = 3.2) that found within our own species today and about half that found between living humans and chimps (25.6/55.0 = 0.47). These results, summarized in Fig. 8.2, show that Neandertals are genetically different from living humans but the difference is not as great as found between humans and chimpanzees. The intermediate results are compatible with the evolutionary time scale—humans and chimps diverged between four and six million years ago, whereas the Neandertals are often considered to have branched off only within the past million years.

If we take these results to indicate that Neandertals were a separate species, we can use standard molecular dating methods to estimate the date of the last common ancestor of the lines leading to Neandertals and living humans by deriving a calibration date from the human-chimpanzee split and then using the derived mutation rate to determine when Neandertals and our ancestors diverged. Using a calibration date of 4–5 million years for the human-chimpanzee split, Krings and colleagues estimated that Neandertals diverged between 550,000 and 690,000 years ago. This range of dates fits nicely with the interpretations of some paleoanthropologists, who suggest a relatively early (Middle Pleistocene) split between the lines leading to Neandertals and our own ancestors. Before declaring the issue settled, however, we

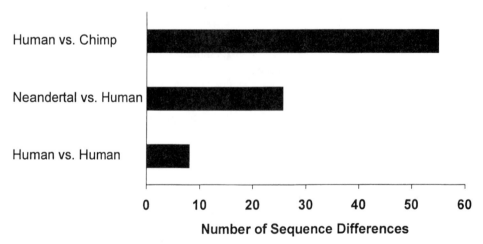

Figure 8.2. *Comparison of the number of sequence differences between the mitochondrial DNA of living humans, chimpanzees, and the Feldhofer Neandertal based on a 333-bp sequence from hypervariable region I (HVRI). (Krings et al. 1997).*

need to remember that these dates apply to a model where we *assume* that the Neandertals actually were a separate species. If this is not the case, then the dates refer only to the history of mitochondrial DNA and not the history of ancient human populations.

The second analysis of the Feldhofer specimen, which added the second hyper-variable region (HVRII) of mitochondrial DNA, provided similar results. They found 12 nucleotide differences out of 340 base pairs (3.5%) when comparing the Feldhofer HVRII sequence with the Anderson reference sequence. They then combined the HVRI and HVRII data for comparison with living humans and living apes (pooling the closely related chimpanzees and bonobos). The combined HVRI-HVRII Feldhofer sequence showed an average of 35.3 differences (range of 29–43) when compared to a sample of 663 humans and an average of 94.1 differences (range of 78–113) when compared to nine chimpanzees and bonobos. By comparison, they found only an average of 93.4 differences (range of 78–113) between humans and apes. They also found an average of 10.9 differences (range of 1–35) between pairs of living humans and an average of 54.8 differences (range of 1–81) between pairs of apes. The results, shown in Fig. 8.3, again show that the difference between the Feldhofer Neandertal and living humans is more than found among living humans but less than the difference between humans and apes. The Neandertal–human difference is also less than that found among the nine apes, but we must remember that the ape comparisons involve two different species (chimpanzees and bonobos). Using a range of 4–5 million years for a human-chimp split for calibration, Krings and colleagues estimated that the common ancestor of Neandertals and living humans lived between 317,000 and 741,000 years ago.

To get a better picture of mtDNA variation within apes, Krings and colleagues compared the human–Neandertal sequence difference with that found within the three subspecies of living chimpanzee: the western (*Pan troglodytes verus*), central (*P. troglodytes troglodytes*), and eastern (*P. troglodytes schweinfurthii*) subspecies. They examined DNA sequence differences for a 312-base pair sequence of HVRI. The average number of DNA differences *within* subspecies was 7.9 for eastern chimps, 14.6 for central chimps, and 21.8 for western chimps. The average

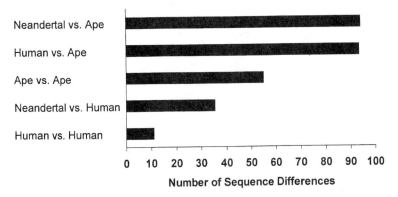

Figure 8.3. *Comparison of the number of sequence differences between the mitochondrial DNA of living humans, apes (chimpanzees and bonobos), and the Feldhofer Neandertal based on pooled data from a 333-bp sequence of hypervariable region I (HVRI) and a 340-bp sequence of hypervariable region II (HVRII). (Krings et al. 1999).*

Neandertal–living human difference (25.6 for this particular 312-base pair sequence) is much larger than any of these values, suggesting that Neandertals and living humans are too different to be considered in the same subspecies, at least when using chimpanzee variation as a guide.

The extraction of mitochondrial DNA from the second Neandertal specimen from Mezmaiskaya Cave supports the results obtained from the Feldhofer specimen. The Mezmaiskaya sequence shows 23 differences when compared to the Anderson reference sequence. The estimated common ancestor for the two Neandertal specimens was estimated as between 151,000 and 352,000 years ago, and the common ancestor of Neandertal and living humans was estimated to have lived between 365,000 and 853,000 years ago.

Many have argued that these studies all confirm the expectation of a replacement model, in which Neandertals and modern humans were different species that diverged from a common ancestor (*Homo heidelbergensis*?) during the Middle Pleistocene. According to this view's proponents, Neandertal DNA is just *too* different to be in the same species as living humans. Although the Neandertal DNA sequences are indeed different, the evolutionary significance of this difference is less clear to me. For one thing, when we compare Neandertal and living human DNA, we are making comparisons across tens of thousands of years. To what extent is Neandertal DNA different because the individual lived a long time ago? Magnus Nordborg addressed this question based on the Feldhofer sequence, noting that there was a relatively high probability (= 0.52) that living human mitochondrial DNA is different because of genetic drift over time.[9] He further notes that the hypothesis of interbreeding between Neandertals and modern human ancestors cannot be rejected, or confirmed, at present. Nordborg notes that even *one* Neandertal sequence similar to living humans would be sufficient to rule out total replacement of the Neandertals. However, we have only so far examined sequences from two Neandertals.

8.3.3 Regional Affinities of Neandertal DNA

Another argument used against a multiregional interpretation of Neandertal evolution is the use of regional comparisons. Neandertal DNA sequences are compared region by region to estimate the average pairwise number of nucleotide differences. The Feldhofer HVRI sequence was compared to living humans from different continental regions. The average number of differences was 28.2 for Europeans, 27.1 for Africans, 27.7 for Asians, 27.4 for Native Americans, and 28.3 for Australians and Oceanians. These comparisons show that the DNA of the Feldhofer Neandertal, who lived in Europe, is not more similar to living Europeans than to other continental groups. There is no significant difference among the regional comparisons, and Feldhofer is genetically equidistant to all regions.

Analysis of the Feldhofer HVRII sequence gave the same story. The average number of differences was 34.4 for Africans, 35.8 for Europeans, and 33.8 for Asians. The Feldhofer sequence is no more similar to one region than to another. The analysis of the Mezmaiskaya DNA sequence produces similar results—an average of 23.1 differences for Africans, 23.3 differences for Asians, and 25.5 differences for Europeans.

There appears to be no significant regional pattern when Neandertal DNA is compared with that of living humans. Some have argued that this lack of pattern supports a replacement model. If all living humans are descended from a single regional population (Africa) during the past 100,000 years, then all living human diversity must have arisen since that time. Because the estimated common ancestor of Neandertals and living humans lived well before this date, the Neandertals are an extinct side branch of humanity that did not contribute to our species' gene pool. We thus expect to see the same pattern that is observed—equidistant similarity of Neandertal DNA with different groups of living humans.

The fact that observed data fit a predicted model is suggestive that a model is correct but does not constitute proof unless alternative models generate different predictions. In the case of regional affinities of Neandertal DNA, the suggestion has been made that the observed equidistant pattern of regional similarity is *not* what is predicted under a multiregional model. According to a broad multiregional interpretation, the Neandertals of Europe were one of a number of regional human populations all interconnected via gene flow. Because multiregional evolution requires isolation by distance, such that most individuals reproduce in, or near, the population they were born into, we would logically expect that, at any given time, most European Neandertals mated with other European Neandertals close by. The logical extension of this pattern of endogamy (mating within one's group) is that living Europeans should be the most similar to the Neandertals. This assumption is stated clearly by Krings and colleagues, who note, "It has been suggested that Neandertals were among the direct ancestors of modern Europeans. European mtDNAs therefore might be expected to have fewer nucleotide differences from the Neandertal mtDNA than mtDNAs from Africa or Asia."[10]

Thus we have two different predictions: 1. Under a replacement model, Neandertal DNA should be equidistant to all geographic regions of living humans. 2. Under a multiregional model, Neandertal DNA should be most similar to the DNA of living Europeans. Because the observed data clearly fit the predictions of the first model, many have suggested that the pattern or regional affinities is proof that a replacement model is correct and the multiregional model is rejected. Although it is possible that the replacement model *is* correct, I disagree with the basic assumption that the multiregional model predicts closest similarity of Neandertals and living Europeans, an assumption that confuses per-generation endogamy with accumulated ancestry over many generations.

The basic question is, What are the genetic effects of gene flow over many generations?[11] This question is usually answered by examining a migration matrix. As described in Chapter 6, this is a table of numbers that reflect the probability of gene flow from one population into another. There are as many rows and columns as there are populations being studied. The entries in the table are the probabilities that a gene in a given population (corresponding to a particular column) came from a given population (corresponding to a particular row). As an example, consider the following migration matrix for two populations, designated A and B.

	A	B
A	0.99	0.03
B	0.01	0.97

The columns represent the current population, and the rows represent one generation earlier. The values in the columns represent the probability of where a gene originated; in this case, the probability of a gene in A having come from A is 0.99, and the probability of a gene in A having come from B is 0.01. Looking at column B, we see that the probability of a gene in B having come from A is 0.03, and the probability of a gene in B having come from B is 0.97. Another way of expressing this matrix is to note that 1% of the genes in A came from B and 3% of the genes in B came from A. It is clear that both populations A and B are endogamous, in that most individuals (and genes) come from the same population. This is a typical pattern in most populations, human or otherwise, and shows the effect of isolation by distance—most individuals choose mates in their own population, or close by.

The trick is figuring out how these rates of gene flow translate into the amount of accumulated ancestry expected over time. At first glance, one might simply take the actual numbers and figure them constant over time; for example, if 99% of population A's genes came from the same population, then we might expect that, over time, 99% of the ancestry of individuals in population A comes from the same population. Actually, it is not that simple. Probabilities multiply over time. The easiest way to figure out the accumulated ancestry over time is to raise the migration matrix to the number of generations. This requires a mathematical method known as matrix multiplication. To figure out the accumulated ancestry after two generations, the matrix above would be raised to the second power, which gives

	A	B
A	0.98	0.06
B	0.02	0.94

Note that the large values became smaller (0.99 to 0.98 and 0.97 to 0.94) and the small values became larger (0.01 to 0.02 and 0.03 to 0.06). After two generations (assuming the rates of gene flow remain constant), population A has 98% of its ancestry from population A and 2% from population B. As time goes on, again assuming the rates remain constant, the numbers change even more. After 10 generations of gene flow, the accumulated ancestry is determined by raising the original migration matrix to the power 10, which gives

	A	B
A	0.92	0.25
B	0.08	0.75

The amount of accumulated ancestry within populations has dropped even more. After 100 generations, the matrix is

	A	B
A	0.75	0.74
B	0.25	0.26

Now, *both* populations A and B have received the majority of their ancestry (75% and 74%) from population A. The amount of accumulated ancestry continues to change slightly until equilibrium is reached with values of

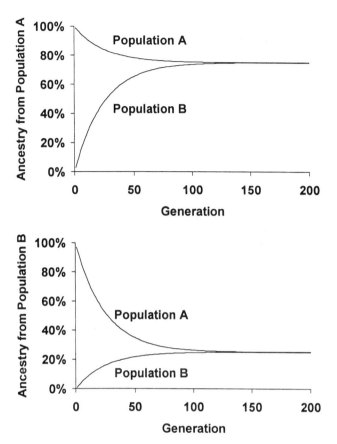

Figure 8.4. *Simulation of accumulated ancestry in two populations. In each generation. population A receives 1% of its genes from population B, whereas population B receives 3% of its genes from population A. Over time, the ancestral contributions to each population are the same—75% from population A and 25% from population B.*

	A	B
A	0.75	0.75
B	0.25	0.25

At equilibrium, the genetic composition of both populations is the same—75% ancestry from population A and 25% ancestry from population B. These changes over time are summarized in Fig. 8.4.

In this simple example, ancestry from population A dominates over time, such that both populations have 75% ancestry from A after 100 or so generations. The reason for this effect is made clear by looking at the difference in the rates of gene flow in the migration matrix. Each generation, population A receives 1% of its genes from population B, whereas population B receives 3% of its genes from population A. The rate of gene flow from A into B is greater than the reverse, so that population A has more overall impact on the final equilibrium frequencies. It is important to note that at equilibrium the ancestral contribution of each population is the same; the ancestry from population A is the same (75%) for both A and B, whereas the ancestry from population B is the same (25%) for both A and B. At equilibrium, we would therefore predict that populations A and B are *both* genetically the same distance from the initial genetic structure of populations A and B.

We can extend these findings to the situation of Neandertal DNA. Of course, we don't have any direct estimates of gene flow to or from the Neandertals, but we can generalize the basic principles to predict that, given enough time, a multiregional model of interregional gene flow would lead to a situation where the genetic distances from Neandertals to different regions of living humans would be the same. In other words, given certain assumptions, *both* replacement and multiregional models predict the same observed outcome. Thus we cannot use regional affinities of DNA *by themselves* to address questions of Neandertal evolution, because the same outcome is predicted under both models. We must rely solely on the magnitude of the difference between Neandertal and living human DNA to address the fate of the Neandertals.

8.3.4   Were Neandertals a Different Subspecies?

The difference between Neandertal and living human mitochondrial DNA sequences is greater than seen within living humans and within living chimpanzee subspecies. This suggests that at the very least the Neandertals were sufficiently genetically different that they and living humans do not belong to the same subspecies. Does this necessarily mean that Neandertals were a different species? Perhaps Neandertals belonged to a *different* subspecies, as many anthropologists have suggested in the past by the placement of the Neandertals in the subspecies *Homo sapiens neanderthalensis* instead of *Homo sapiens sapiens*, the subspecies to which all living humans belong. The past placement of Neandertals as a separate subspecies was a way of noting their physical differences from living modern humans for certain anatomical features while at the same time acknowledging their close overall similarity in other features, such as a large brain size.

Krings and colleagues performed several comparisons *between* chimpanzee subspecies that have some relevance to this question. Based on a 312-base pair sequence for HVRI, they found an average of 19.7 DNA differences between the central and eastern subspecies, 33.0 differences between western and eastern subspecies, and 36.2 differences between western and central subspecies. The average difference between living humans and the Neandertal specimen (= 25.6) is less than between two out of three chimpanzee subspecies comparisons (see Fig. 8.5). These findings suggest that Neandertals and living humans *could* belong to different subspecies within the same species, especially if we consider that the chimpanzee comparisons are all made at one point in evolutionary time (the present), whereas the Neandertal-living human comparison encompasses between 35,000 and 70,000 years, depending on the exact date of the Feldhofer specimen.

8.4   SUMMARY

The extraction of Neandertal DNA represents a marvelous technological breakthrough in the study of ancient DNA. Unlike the reconstructions discussed in earlier chapters, derived from inferences based on the genetics of living humans, we now

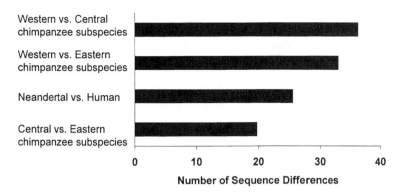

Figure 8.5. *Comparison of the number of sequence differences between the mitochondrial DNA of living humans and Neandertals compared to differences between chimpanzee subspecies based on a 312-bp sequence from hypervariable region I (HVRI). (Krings et al. 1999).*

have some data on the genome of earlier hominids. Mitochondrial DNA has so far been extracted from two Neandertal specimens, Feldhofer and Mezmaiskaya, and the scientific community awaits further extractions.

Although there are only two specimens analyzed to date, the results seem clear— the mitochondrial DNA sequences are quite different from that of living humans. Whether this difference translates into the widely echoed sentiment among the media—that these data prove conclusively that Neandertals were an extinct species that was replaced—remains to be seen. There is a large difference between the sequences of the two Neandertal specimens and living humans. This large a difference exceeds the range of variation typically seen within living humans and within subspecies of chimpanzees. As such, it seems unlikely that we could consider Neandertals and living humans *within the same subspecies*. On the other hand, the difference between Neandertals and living humans fits comfortably within the range of difference found when comparing subspecies of chimpanzees to each other.

On the basis of ancient DNA analyses, there are two possibilities—Neandertals were a separate species, or Neandertals were a subspecies within a single human species. Both outcomes imply some degree of reproductive isolation, but the former implies a lack of genetic continuity with our ancestors and the next task is to determine why replacement of one species by another took place. The latter leaves open the possibility that this isolation was not sufficient to produce barriers to gene flow. If so, then gene flow between modern humans and Neandertals was possible, and they are *part* of our ancestry, and the next task is to determine how much. The fate of the Neandertals, whether through extinction/replacement or genetic absorption, remains unclear.

Of course, all of these analyses require that the mtDNA is not affected strongly by natural selection. Because a person's mtDNA is inherited as a single unit, without recombination, selection for any part of the mitochondrial genome could result in new mitochondrial haplotypes being spread throughout a species. If this is the case, then all of our reconstructions of history may be showing the history of natural selection for mtDNA and not the history of the populations. One could therefore postulate that one reason for the genetic difference between Neandertal and living human mtDNA is natural selection. Although our observations of mtDNA diversity are generally consistent with neutral models, we should always keep the possibility of selection in mind and continue to pursue research that will help answer this question.

A major problem in focusing on a single trait, such as mitochondrial DNA, is that even under neutrality we are seeing the history of essentially a single locus. Because of the statistical fluctuation in the evolutionary process, we should not expect any single trait to show the true underlying pattern of history. It would be useful to have evidence from other genetic traits, including nuclear DNA. We may be facing a limit to current technology; because nuclear DNA is less frequent than mitochondrial DNA, we may not be able to extract enough of the former for analysis. This is certainly the case with the Feldhofer specimen, where analysis shows little hope of extracting nuclear DNA. The situation might be different for the better-preserved Mezmaiskaya specimen; only time will tell.

Resolution of the evolutionary status of the European Neandertals requires more than comparison of Neandertals with living humans. The debate might be settled if we were able to compare Neandertal and living human DNA sequences with ancient DNA from European fossils that are undeniably modern human but that lived close to the time of the Neandertals, such as the Cro-Magnon specimens. To date, we do not have such sequences. If they can be extracted, the comparisons might shed further light on the debate. If, for example, a small difference between early modern and living human DNA would strongly favor a model of Neandertal replacement. On the other hand, a small difference between early modern and Neandertal DNA would suggest that such forms were genetically related, and would lend support to a multiregional interpretation. An intermediate difference between early modern and Neandertal DNA would be more problematic, perhaps not favoring either model to the exclusion of the other.

Further studies of ancient DNA may provide us with the needed information to address replacement versus continuity in Europe, something that we can possibly look forward to as the science of ancient DNA analysis continues to develop. For the moment, I suggest that the available data support some degree of genetic isolation of the Neandertals, although the question of whether this isolation reflects speciation is still in doubt. Finally, it is worthwhile repeating that any answer to the question of *Neandertal* evolution does not tell use what happened in the rest of the Old World. Replacement in one region does not mean that replacement occurred elsewhere.

8.5 POSTSCRIPT

One of the fascinating, although often frustrating, aspects of genetic research on modern human origins is the pace of new discoveries. In October 2000, while this book was already in production, the extraction of mitochondrial DNA from a *third* Neandertal specimen was published.[12] Matthias Krings and colleagues extracted 357 bp of hypervariable region I and 288 bp of hypervariable region II from the Vindija 75 Neandertal fossil from Vindija Cave, Croatia. The results were similar to those from the analyses of the Feldhofer and Mezmaiskaya specimens. The authors note that these data suggest that Neandertals did not contribute any mtDNA to living humans, but that this does not rule out the possibility that interbreeding took place between Neandertals and modern humans. They also show that three Neandertal specimens are very similar, suggesting that, like living humans, diversity in Neandertal mtDNA was relatively low. Given a sample size of 3, further samples are needed to confirm this hypothesis. The new sequence from Vindija does not alter any of the suggestions I have made in this chapter.

As of January 2001, we still do not have any mtDNA sequences from European fossils that are anatomically modern. However, some evidence has emerged across the world that has some implications for the Neandertal studies. Gregory Adcock and colleagues published a paper in January 2001 in the *Proceedings of the National Academy of Sciences*[13] describing the extraction of mitochondrial DNA sequences from a series of Australian fossils, including the Lake Mungo 3 (LM3) specimen thought to date to roughly 60,000 years. This specimen is particularly important because of its age and the fact that it is anatomically modern and shows morphological continuity with living Australian aborigines. The mtDNA sequence of LM3 is divergent, however, and does not resemble that of living Australians. The full implications of this analysis are still being considered, but one striking observation is that the presence of an extinct mtDNA sequence in an ancient *anatomically modern* fossil provides an excellent example of the extinction of mtDNA sequences over time, as predicted by many population genetic models. This finding does not bear directly on the Neandertal issue, but does suggest that if mtDNA sequences from that far back in the past can become extinct, then the absence of Neandertal mtDNA in living humans does not necessarily support a replacement model.[14] Just as LM3 is ancestral to living humans, it is also possible that Neandertals contributed some ancestry to living humans, even though their mtDNA has been lost.

Chapter **9**

Putting the Pieces Together

The previous five chapters have reviewed a wide range of genetic data, much of which has often been cited as support for an African replacement model of modern human origins. In each case, closer examination suggests the picture might not be that simple. Although the data are certainly compatible in most cases with a replacement model, it is clear that they are also compatible with one or more multiregional interpretations (again, using the word "multiregional" in the broad sense of more than one geographic region and not just the narrower interpretation of the regional coalescence model).

At this point, it is useful to review the main points of the previous chapters and to consider them in relation to the fossil evidence. Questions about modern human origins are not going to be completely answered by an exclusive focus on either genetic or fossil data, but through a synthesis of both. This is of course a difficult process, given the fact that it is hard enough to become conversant with one area, let alone two. The problem is sometimes exacerbated by the different mindsets of geneticists and paleoanthropologists. Different fields use different methods and often have different ways of looking at the same evidence. Add to this the tendency of many to elevate their own specialization above others, and we see that interaction between the different disciplines is difficult at best.

Still, anthropology is nothing if not interdisciplinary in nature. As practiced in the United States, the study of anthropology deals with both biology and culture and has roots in the natural and social sciences as well as the humanities. It is increasingly difficult to maintain this holistic orientation in an age of ever-increasing specialization, but I maintain that there are some benefits as well. In the case of modern human origins, most researchers focus their efforts on one small piece of the puzzle, be it genetic, paleontological, or archaeological. As time goes on, it is common for all of us to further narrow the scope of our efforts, becoming involved with (obsessed with?) tinier and tinier pieces of the puzzle. I am familiar with this trap in my own work, where I have spent days on end dealing with esoteric details of obscure equations only to find that a simpler approach existed or one better suited for the real world. Sometimes this focus leads down familiar alleys, where in a burst of effort you rediscover something already known and you have literally reinvented the wheel (I remember as a graduate student rediscovering Hardy–Weinberg equilibrium late one night—an intellectually satisfying experience, but not something one can write a dissertation on).

Specialization and increasing focus is part of the scientific and educational process. You start out as an undergraduate with a broad survey of a particular field, moving into more detail as you begin graduate school. By the time you have begun working on your doctoral dissertation, your efforts have become more and more focused, and your research area becomes (to you) the most central thing in your intellectual universe. Later in your career, you begin to move back out of the hole you have dug and hopefully broaden your interests, focusing more and more on the *meaning* of your work in relationship to the larger picture, and asking "What does this mean?" I personally find teaching undergraduates a valuable tool for making this transition, because they will ask you literally, "What does it mean?"

In the case of studying modern human origins, people tend down one of several paths, including paleoanthropology, archaeology, population genetics, or molecular genetics (among others). At some point, however, we all need to talk to one another (and not *at* one another). This is difficult because we have just enough in common sometimes to get into trouble, assuming that everyone uses the same terms in the same ways (e.g., the debate over the underlying predictions of different models). In anthropology, an additional problem that sometimes surfaces is the long-standing feud between paleoanthropology and genetics dating back to the debate in the late 1960s over the timing of the ape–human split. Although the dust has more or less settled on that debate, many of the feelings remain, and we continue to see examples of geneticists scoffing at the fossil record and paleoanthropologists dismissing input from geneticists. The reassuring thing is that, at least from my own perspective, this cross-disciplinary sniping is fading to some extent (or I am becoming more of an optimist).

It is easy for me to advocate that paleoanthropologists and geneticists each consider the evidence from each other's field, but this is more difficult to do in practice. It is a daunting task to consider a review of another subdiscipline in your field when you are not an expert in it. This is particularly scary at the midpoint of your career, when the natural cockiness of graduate school has worn off and you discover how many things you don't understand about your own subdiscipline! Nonetheless, I think the effort is worth it.

9.1 SUMMARY OF THE GENETIC EVIDENCE

What can we say about patterns of genetic variation observed in living human populations (and now from the Neandertal DNA sequences)?

- Many DNA sequences show an African root (coalescence in Africa). Some sequences also show deep non-African branches.
- Coalescent dates for mtDNA are generally around 200,000 years. Dates from Y chromosome DNA are a bit younger. Dates from nuclear DNA are older.

- Coalescence dates tell us about the history of specific genes and DNA sequences and not the history of populations. Coalescence dates for mtDNA and Y chromosome polymorphisms suggest a level of genetic drift compatible with a relatively small species effective population size over at least the last several hundred thousand years.
- Genetic diversity is highest in sub-Saharan African populations, suggesting that the long-term average effective population size in Africa was larger. Non-African populations would therefore have experienced greater genetic drift.
- Patterns of genetic distance among regional populations show that sub-Saharan Africa tends to be the most different, suggesting variation in gene flow in general and, specifically, more gene flow out of Africa than into Africa.
- A relatively small average number of migrants per generation between geographic regions can explain patterns of genetic distance and diversity. It is not clear *how* this gene flow occurred (isolation by distance versus admixture with dispersing populations) or the time depth of the migration.
- The species effective size of our ancestors is equivalent to about 10,000 adults averaged over the past several hundred thousand years. There is evidence for a Pleistocene population explosion resulting in a rapid increase in species effective size on the order of 40,000–50,000 years ago. Given this expansion, the species effective size *before* the expansion is likely to have been even smaller, perhaps only a few thousand. It is not clear whether the expansion was recovery from a bottleneck of short duration or whether effective species size was small for the past one to two million years.
- Genetic data provide us with valuable information about species *effective* size, but the relationship of these measures with species *census* size is less clear. In particular, we don't know if the low species effective size implies a low census size or a much larger census size under conditions of the frequent extinction and recolonization of local populations.
- Initial extractions of mitochondrial DNA from Neandertal fossils shows significant differences relative to living humans, although the interpretation is less clear. Are the sequences different because they lived a long time ago, because they were a relatively isolated subspecies within *Homo sapiens*, or because they were a separate species that diverged from our common ancestor over half a million years ago?

All of these points are compatible with the African replacement model. It has become almost an everyday occurrence to read a news item proclaiming this compatibility as proof of the replacement model. In some circles, this proof is now so thoroughly established that the debate over modern human origins has been declared over, with the winner being the African replacement model. I was watching a television documentary on human evolution last year and was struck by its dogmatic nature and kept waiting for alternative viewpoints to be aired.

It is certainly possible that the African replacement model is correct. My reservations suggest more caution, because we must first rule out that multiregional alternatives, in the broad sense, are not also compatible with the data. I find that for every point listed above there exists a viable alternative to a replacement model. In my view, the genetic data *are* valuable and *do* tell us something about our past, but not in a simple direct manner that equates to a phylogenetic interpretation. What works for comparing different living species does not work the same way when comparing different groups of living humans, all of which belong to the same species. Phylogenetic models that rely on a mutation-drift balance in isolated lineages cannot be applied to a set of populations that have experienced gene flow. Continental groups (what some folks call "races") cannot be treated as independent evolutionary entities. Because we have only one living human species, many of the models used by evolutionary biologists studying other species cannot be used, and we need to develop alternative methods.

If not phylogenetic history, then what *do* the genetic data tell us about our past? In my view, the primary and overriding signal is demographic—genetics tells us about the past in terms of population size (drift) and migration (gene flow). My own view is that much of the genetic evidence can be interpreted by two basic findings, both concerning population size: 1. Our long-term effective species size is small (~10,000) and 2. Throughout much of our prehistory, effective population size was larger in Africa. I find the second point particularly revealing in terms of modern human origins. My view is that our species is, from a demographic perspective, *mostly out of Africa*, meaning that the larger effective size of Africa has had the largest regional impact on genetic variation in our species. As is outlined below, this view can be expressed in terms of both replacement and multiregional models, although I favor the latter. Before outlining this model in detail, it is useful to briefly reconsider the fossil evidence for the modern human origins debate. Because the genetic data by themselves cannot resolve the debate, we need to examine the actual physical record of past human evolution.

9.2 THE FOSSIL RECORD—A POPULATION GENETICIST'S VIEW

I have always been fascinated with the human fossil record and attempt to keep conversant with the latest finds and models dealing with our evolutionary history. Because I am not an anatomist by training, reviewing the literature is often rough going (as I imagine some of the genetics literature is to paleoanthropologists). I often read the latest journal articles with one hand on the journal and the other on a reference text, trying desperately to remember the anatomy learned in graduate school over 20 years ago. I also spend considerable time examining casts and pictures of relevant specimens trying to understand at a rudimentary level the complex morphology that my paleoanthropology colleagues see so easily. My holistic training in anthropology suggests that the effort is worth it (although my attitude does not guarantee success by any stretch of the imagination!).

Given my own expertise (and related biases), I find myself drawn primarily to the level of analysis, particularly when these studies relate more or less to the types of research that I do. In the broad sense, there are some similarities that I can latch onto for understanding, primarily general questions regarding evolutionary history. At this general level, the questions concern the relationship of populations over time and space, something I am more familiar with from my other work on much more recent population history (e.g., Ireland over the past thousand years).

There are disagreements over the specific interpretations of the human fossil record. To some, the fossil record is clearly a record of a recent African origin with subsequent replacement. To others, the same fossil record provides a picture of a recent African origin of modern humans, but within a multiregional context. Still others see the transition from archaic to modern humans as part of a regional coalescence. Which view(s) are more likely to be correct, and how do they fit with the genetic evidence?

My own biases predispose me to look favorably upon analyses that deal with variation in the broadest sense, comparing samples from different times and places in an effort to reconstruct our evolutionary history. I view variation within populations as normal, such that the concept of a "typical" specimen is sometimes difficult to deal with except in an abstract sense. When I see two fossil specimens, my natural inclination is to focus on the similarities and not the differences. This makes me tend to be a "lumper" for the most part, and again this is in part a reflection of my training. However, the increasing evidence for a very wide range of variation among our early ancestors, the australopithecines, suggests that there are times in evolutionary history that frequent speciation was the rule, and maintaining my "lumper" perspective at all times might be inappropriate. To me, each case might be different, and I cannot state a priori that "lumping" or "splitting" is the best approach to take when considering the variation with the fossil record for the genus *Homo*.

When considering the variation in *Homo* across time and space, I am less interested in arguments about the number of species or what to call them than I am in the distribution of variation. How do fossil samples from different times and places compare with one another, and how do these patterns fit our predictions regarding the different models of modern human origins? I see three main sources of relevant information: 1. The distribution in time and space of archaic and modern fossils, 2. The similarity of recent modern humans (including living humans) to earlier archaic and modern samples, and 3. Patterns of regional continuity.

9.2.1   The Distribution of Archaic and Modern Fossils

An obvious primary step in evaluating the fossil record would be to consider the distribution of archaic and modern human fossils across time and space. When did modern humans first appear in different geographic regions? This is not always that simple a question given the fact that there is often disagreement over how to classify any given specimen, or even whether the attempt to partition variation into "archaic" and "modern" components makes any sense at all. Even if one agrees that such partitioning is possible, we must deal with the fact that the fossils do not always show an "either-or" appearance, and we must consider variation within the general categories of "archaic" and "modern" and the fact that the ranges of variation overlap. Thus we have to deal with fossils that might need further qualification, such as "early modern" or "intermediate."

Some paleoanthropologists suggest that modern humans did not arise in any single place as a population but, instead, modernity gradually evolved through the coalescence of different modern traits appearing in different places over time.[1] This view, which I termed "regional coalescence" in Chapter 3, states that there is no single origin point. If this interpretation of the fossil record is correct, the African replacement model and the primary African origin models are wrong because they both propose the origin of modern traits in a single region followed by either replacement or genetic exchange with other human populations.

A growing number of paleoanthropologists feel differently (although consensus of opinion is not useful as a measure of validity). Many suggest that there was a recent African origin, such that modern human traits begin to appear in Africa earlier than in other regions. According to this interpretation, the earliest evidence for anatomically modern humans is from sites in Africa (such as Border Cave, Omo, and Klasies River Mouth) dating back between 100,000 and 130,000 years ago. Early modern humans are found slightly later in the Middle East, dating back roughly 90,000 years, and in Australia by at least 50,000 years ago. The appearance of modern humans in Europe is often suggested to have taken place roughly 30,000 to 40,000 years ago, with the younger dates associated with Western Europe.[2]

Even with caveats and qualifications, the geographic pattern seems to fit an African replacement model, which predicts modern humans arising first in Africa and then spreading to the Middle East and then to more distance locations later in time. The lag in time makes sense geographically because it would take time for an expanding species to move across the entire Old World—replacement is not likely to have occurred instantaneously. Given agreement between observed and predicted patterns of geographic distribution over time, it is tempting to use this correspondence as proof of African replacement. However, the same pattern is also predicted by primary African origin models that combine an initial origin of modern traits in Africa with gene flow outside of Africa. According to this view, the geographic distribution of modern fossils shows the spread of *genes* within a species and not the spread of a new species. Furthermore, if this interpretation is correct, the geographic distribution does not help us distinguish between gene flow due to migration between local populations or admixture with populations moving out of Africa. Models of gene flow also predict a temporal lag, because it takes time for people or genes to move across space, particularly in hunting-gathering times. The question remains—are we seeing evidence of the movement of populations replacing other populations, the movement of populations mixing with other populations, or the movement of genes from population to population? To make things more complicated, we should also consider the possibility that these may not be mutually exclusive; for example, we might be seeing a complicated picture of gene flow through isolation by distance in some regions, admixture with dispersing populations in others, and perhaps even replacement in some places.

9.2.2 Similarity of Recent Moderns to Earlier Archaics and Moderns

A number of studies have taken the analysis of the geographic distribution to the next level by comparing the degree of biological similarity between fossil samples from different times and places. Such studies typically look at the average patterns of biological distance between samples based on a number of different anatomical traits, including both measurement data (e.g., head length) and the presence/absence of different traits. The underlying idea is to use as many traits as possible to minimize statistical error and evolutionary fluctuations (due to drift) and get an estimate of the average degree of dissimilarity. Which earlier samples tend to look more similar to later modern humans? Although the studies differ in terms of different samples and traits used for analysis, the results tend to be similar in finding the closest similarity to earlier African and Middle Eastern samples.[3]

This similarity is most often interpreted as support for African replacement. If all recent modern humans (within the past 30,000 or so years) are descended from Africa within the past 100,000 years ago, then comparisons across time should show the greatest similarity with those earlier samples (assuming subsequent gene flow and genetic drift has not altered the underlying history). They do. Again, the observed data are compatible with the expectations of an African replacement model, but once again, this does not constitute proof. We must first ask what the expectation is under a multiregional model and determine that the data are not compatible with it before ruling in favor of the African replacement model.

Some have suggested that under a multiregional model the smallest biological distances across time periods should be within the same geographic region. Thus, if a multiregional model is correct, then recent modern Europeans should more closely resemble earlier archaic Europeans (the Neandertals) than earlier fossil samples in Africa or Asia. This reasoning was discussed in general in Chapter 3 and more specifically in Chapter 8 regarding Neandertal DNA. The underlying idea is that regional endogamy predicts closer relationships within regions over time than between regions over time. A comparison of the predictions of African replacement and multiregional evolution is illustrated schematically in Fig. 9.1. Here, we consider fossils from two geographic regions, A and B, which are both sampled at time periods 1 and 2 (the second time period being more recent). We are interested in comparisons across the two time periods under different origin models using lines to indicate the comparisons that should show the greatest similarity across time. Figure 9.1a shows the prediction from a replacement model in which all living humans arise from region A and earlier humans from region B become extinct. Lines are drawn from sample A1 to both of the more recent samples, indicating that the later samples should be more similar to A1 than to B1. Figure 9.1b shows a commonly stated prediction of multiregional models; because most genes are shared within geographic regions, the greatest similarity (as indicated by the lines) should be within regions. Thus we would expect sample A2 to be more similar to sample A1 than to sample B1 and sample B2 to be more similar to sample B1 than to sample A2. Given these predictions, the finding that recent modern humans more closely resemble earlier samples in Africa than elsewhere is taken as evidence for a replacement model. The problem with this inference is the predictive model for multiregional evolution (Fig. 9.1b) is wrong. Given what we know about ancient population dynamics from genetic data, the correct predictive model is Fig. 9.1c, which is identical to the prediction of the replacement model.

How can this be? As shown in Chapter 8, the rates of migration in any given generation are not equivalent to the amount of accumulated ancestry over many generations. Here, I use an example published elsewhere[4] to consider further the impact of regional differences in population size. Imagine two populations, A and B, where A consists of 4000 adults and B consists of 1000 adults. Let us further imagine that these two populations exchange 10 mates each generation. We can illustrate this model with a matrix of migrant numbers following the same

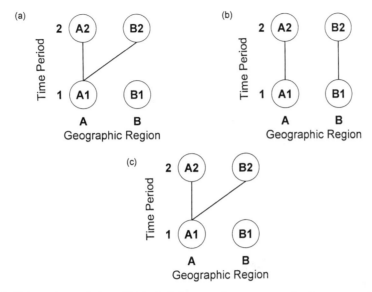

Figure 9.1. *Schematic prediction of biological distances between samples from two regions (A and B) over two time periods (1 and 2). Each population is coded by its region and time period; thus A1 represents a sample from region A in time period 1. Solid lines indicate the smallest biological distances across time predicted by a given origin model. a. Replacement model. All populations in time period 2 derive from population A1 such that the shortest distance across time is to A1. b. Multiregional evolution (false). Similarity is greatest within regions across time. c. Multiregional evolution (true, based on differential effective population size). Both A1 and B1 contribute genetically to both A2 and B2, but the greatest impact (and shortest distance) is to A1 because population A is larger and contributes more to accumulated ancestry over many generations. Note that (a) and (c) generate the same predictions.*

method discussed in Chapter 8—the rows represent parents and the columns represent their offspring:

	A	B
A	3990	10
B	10	990

Each generation, 10 people in population A came from population B and the remaining 3990 people were born to parents who lived in A. Following standard methods, we divide each column by the number of people in each population to get a matrix of migration probabilities:

	A	B
A	0.9975	0.0100
B	0.0025	0.9900

This is a clear case of endogamy—99.75% of the genes in population A came from A, and 99.00% of the genes in population B came from B. Given the low rates of gene flow between A and B, we might expect that the long-term picture of accumulated ancestry would give rise to the prediction shown in Fig. 9.1b. Not so. The accumulated ancestry after only 200 generations is

	A	B
A	0.8162	0.7354
B	0.1838	0.2646

By this point, 18% of population A's ancestry came from population B, whereas the majority of population B's ancestry (74%) came from A. In terms of Fig. 9.1, by this point in time we would expect sample B2 to be more similar to A1 than to A2. These numbers will continue to change until an equilibrium is reached of

	A	B
A	0.8000	0.8000
B	0.2000	0.2000

At equilibrium the ancestry of *both* populations A and B is 80% from A and 20% from B. These final figures are not unexpected—they represent the relative size of each population:

$$\text{Relative size of population A} = \frac{4,000}{4,000+1,000} = 0.8$$

$$\text{Relative size of population B} = \frac{1,000}{4,000+1,000} = 0.2$$

Under a model of symmetric migrant numbers, the eventual ancestry is proportional to effective population size.[5]

What does this mean in terms of the modern human origins debate? Given the genetic evidence for larger effective population size of Africa, we expect that over time Africa would have the greatest genetic impact and recent modern humans will more closely resemble earlier Africans on average than other regions. This is what many analyses of the fossil record have shown, but in this case, the prediction was derived from a multiregional model rather than replacement out of Africa. As an analogy, imagine mixing four pints of red paint with one pint of yellow paint. The resulting mixture is made up of contributions of both colors, but in different amounts. The mixed paint will be more similar to a can of red paint than a can of yellow paint. The mixture is mostly out of the red can, and analogously we can consider a model of modern human origins that describes our species as "mostly out of Africa." This observation corrects earlier attempts to model multiregional origins but introduces another problem. If both the African replacement model and a multiregional model with a larger African population predict the same thing and are both in agreement with the data, then how can we tell them apart?

9.2.3 Regional Continuity

As discussed in Chapter 3, a key prediction of regional continuity is the similarity of traits over time *within* regions. As I noted in Chapter 3, the concept of regional continuity appears at first to be at odds with the results above by suggesting greater affinity of recent modern humans to earlier non-African humans. Regional continuity *does* predict greater similarity over time within regions (Fig. 9.1b), which has already been ruled out! The paradox is resolved by noting that, under a model of gene flow balanced by genetic drift, *most* traits will not show continuity, but *some* will (refer back to Fig. 3.11 and related discussion). The *average* biological distance over many traits will show the expected pattern of African affinities but some *individual* traits will maintain regional continuity because of drift.[6] Under a multiregional model, we would predict that some traits will show continuity, but most will not.

Anthropologist Marta Lahr has performed a thorough analysis of continuity traits in crania from Asia and Australasia.[7] She found that most traits (63%) did not show any regional pattern or showed higher frequencies in other regions. The low incidence of traits showing continuity within a region, and the fact that no traits were found exclusively within any given region, led her to conclude that there was no evidence for regional continuity or a multiregional model. This conclusion rests upon a view of multiregional evolution requiring strong and exclusive continuity, which I believe is in error. Given a model of gene flow balanced by genetic drift, most traits in non-African populations are *not* expected to show continuity because of gene flow moving the allele frequencies toward those of Africa. This gene flow is countered in some genes by random genetic drift, which can maintain high frequencies of an allele in the presence of gene flow (as was illustrated in Fig. 3.11). The important point is that *some* traits will show continuity because of drift. To me, the fact that Lahr found a minority of traits suggesting continuity is perfectly consistent with a multiregional interpretation.

Regional continuity has been claimed by some paleoanthropologists across the Old World, with particular emphasis on East Asia and Australasia.[8] To some, the situation in Europe is less clear, although even here, where the differences between moderns and archaics (Neandertals) seem most pronounced, a number of studies have argued for continuity over time. Taken collectively, I believe studies of the fossil record have made a strong case for some continuity outside of Africa, therefore ruling out a complete replacement model (but not necessarily a model combining a recent African origin with gene flow outside of Africa).

There are several other points to consider before accepting this conclusion. First, we must confine our observations to neutral traits because things change when natural selection is operating. Natural selection can maintain high frequencies of a regional trait but can also mimic the effects of regional continuity due to common ancestry. Imagine, for example, an adaptive mutation that appears in a given region and is selected for, resulting in a higher frequency compared to other environments. Imagine further that this population becomes extinct and is replaced by another species that has a similar mutation that is also selected for because of the same adaptive reason. The trait will reemerge in this region, even though the first population died out. In this case, regional continuity would be due to adaptive convergence and not the shared ancestry required by a multiregional model. As in any study of population history, we need to exclude traits that are obviously under natural selection, particularly if the selection varies across environments (as would be the case if we wanted to use skin color to reconstruct population history, leading to a conclusion that Australian aborigines are more closely related to sub-Saharan Africans). Many of the traits analyzed in studies of regional continuity appear to have little if any adaptive significance and likely represent minor anatomical variants that originally become prevalent due to genetic drift. However, we should always guard against the possibility that our results may be biased by potential natural selection.

Another potential problem is regional continuity being mimicked by the retention of primitive traits (known as "plesiomorphic" traits by evolutionary biologists). To illustrate this, let us imagine a trait that existed in low frequency in Africa two million years ago but drifted to a high frequency in populations dispersing into Asia. Now, let us further imagine a later speciation in Africa followed by dispersal into Asia replacing the previous Asian hominids. If replacement occurs, then the Asian regional trait should disappear and any persistence would be an indication of continuity in ancestry within Asia. However, suppose that the trait persisted in Africa in low frequency and then drifted *again* to a high frequency among the new migrants into Asia. This pattern would reestablish the trait in Asia through the action of dispersal with drift occurring twice in the same manner. If so, then the persistence of a regional trait could be a chance event and falsely mimic actual genetic continuity within the region. All that this model requires is some persistent variation within the parental species (or recurrent mutation). If we add in variation in observed frequencies because of sampling error from a relatively sparse fossil record, is it a possibility that our signal of continuity is illusory? This seems unlikely, particularly if we extend this process to more than one trait. Although one trait could conceivably reappear, it is very unlikely that the same *set* of characteristics would do so.

It is useful to also examine specifics about changes in regional continuity traits over time. One excellent example is the work done by Dave Frayer and Milford Wolpoff on Neandertal traits.[9] The prevalence of unique Neandertals traits (or very common in Neandertals and rare in other hominids) was examined in the earliest post-Neandertal modern fossils and in living Europeans. Because these traits are absent or very rare in non-Neandertal samples, they should not be present in post-Neandertal samples. If instead there was genetic continuity and not complete replacement, then these traits could persist. One of these traits, the frequency of a particular type of nerve opening on the lower jaw, is considered a unique trait found in many Neandertals but not elsewhere. The frequency of this trait decreases from 53% in the Neandertal sample to 18% in the earliest post-Neandertal modern sample, and further decreases to 1% in living Europeans. The decrease over time fits a model of continued gene flow with the rest of the species outside of Europe (that did not have the trait). A reduction in the frequency of continuity traits is also expected if we take the genetic evidence of Late Pleistocene population growth into account. As effective population size increases, the balance between genetic drift and gene flow changes; specifically, further genetic drift is minimized and gene flow has a greater influence on allele frequency differences. Without continued strong drift to maintain continuity, this neutral trait would decrease over time because of gene flow from outside of Europe, exactly the pattern observed.

Other traits show the same reduction in Neandertal features over time. The suprainiac fossa (an anatomic feature of the upper rear of a skull) is found in 96% of Neandertal specimens, 39% of the earliest post-Neandertal moderns, and 2% of living Europeans. Most of the Neandertal traits were found in the earliest post-Neandertal moderns, and a number of these continued to persist in living

Europeans, although at reduced frequency. These results suggest *some* genetic input from Neandertal to modern humans. This view has recently been supported by a discovery made by Cidália Duarte and colleagues of the burial of a child who lived 24,500 years ago in Portugal.[10] The skeleton shows a mixture of Neandertal and early modern human characteristics, as expected during this time period under a model of gene flow between Neandertals and moderns but not expected under a model of complete replacement out of Africa.

It appears that Neandertals made some genetic contribution into later European populations, although the exact level might have been relatively low, particularly given a larger African population and subsequent genetic impact. It might be argued that continued gene flow from Africa into Europe resulted in the African component "swamping" that of the Neandertals, effectively resulting in a "genetic replacement." In my view, this is still a multiregional model incorporating gene flow from different regions within a single evolving species. Even if the Neandertal contribution was very small, it is still a evolutionary process fundamentally different from speciation and complete replacement.

9.3 "MOSTLY OUT OF AFRICA"

Scientific research often feels to me like working on a jigsaw puzzle picked up at a yard sale and missing some of the pieces, having to figure how not only the pieces fit but make inferences about what is missing. As new evidence accumulates, some views are confirmed and others are rejected. This is the normal mode of operation in science, using accumulating data and new methods to invent, refine, or reject hypotheses. I think the search for modern human origins is particularly difficult because it is a relatively new field. All of the fossil record has only been discovered in the past 150 years, and much of the genetic evidence dates back only a few years. During the same time, evolutionary science has grown from the contributions of Darwin and others through the development of Mendelian and population genetics to today's rapid development of molecular genetics. Although working in such a young and developing field is exciting, it is also frightening because the knowledge base changes so rapidly. Because there is still so much we do not know about the fossil record and the genetics of our species, it is very likely that some of my own thoughts will soon be changed or rejected (perhaps by the time you read this!). Consequently, I am not comfortable declaring one model to be *the* final answer. This does not mean that all are equally viable in my view. Instead of final statements, I prefer to discuss various models in terms of relative probabilities.

I feel that the African replacement model has a relatively low probability of being correct, in the sense of an origins model in which *Homo sapiens* emerged as a new and reproductively isolated species within the past 200,000 years. My conclusion is biased by my views on variation in general and my views on human evolution specifically but is also based on the findings of some non-African influence from some genes and DNA sequences as well as the fossil record for regional continuity. However, I agree with those paleoanthropologists and geneticists advocating a recent African origin of modern *traits*, in large part because of my view that the genetic evidence supports a larger average long-term effective population size for

Africa and the evidence for early moderns appearing in or near Africa. In this sense, my views are similar to those of paleoanthropologists Gunter Bräuer and Fred Smith as well as the views of geneticists Lynn Jorde, Michael Bamshad, and Alan Rogers, who recently wrote

> Although the majority of evidence appears to favor an African origin of modern humans, it is less clear that these modern humans completely replaced non-African populations. There is now limited evidence . . . for ancient non-African contributions to the present-day gene pool. Thus, neither the African replacement model (in its pure form) nor the multiregional model is fully consistent with the data. An African origin, with some mixing of populations, appears to be the most likely possibility.[11]

I differ from Jorde and colleagues in how I label this conclusion. I suggest that this *is* a multiregional model in the broad sense of involving genetic input from more than one geographic region within a single evolving species.

My interpretation of the genetic and fossil evidence is that our ancestry over the past several hundred thousand years is *mostly, but not exclusively out of Africa.* Humans in Africa are viewed as an evolutionary "core" population that was, on average, larger for much of our prehistory, with other regions being on the periphery. This view is similar to that proposed by Alan Thorne and Milford Wolpoff's view of "center and edge," in which gene flow is predominately from a species' evolutionary center (Africa) into peripheral populations on the edge (Europe, Asia, Australasia).[12] Larger effective population size in Africa predicts the observations we make today from the genetics of living human populations. Diversity is greater in Africa than in other regions, where smaller effective size would lead to increased genetic drift. This also means that DNA coalescence would be more recent outside of Africa for the most part, and on average, the deepest branches of a gene tree would be African. Higher rates of gene flow out of Africa than into Africa would replicate the pattern of genetic distances among regions seen today. I also suggest that larger effective population size in Africa with gene flow out of Africa into regions with higher levels of genetic drift also explains the pattern found in some DNA sequences where there are more haplotypes in Africa and diversity outside of Africa represents a subset of African diversity.

How does all of this translate into an evolutionary model for the evolution of the genus *Homo* over the past two million years? Although we still lack many details, I suggest the following scenario. Early humans (*Homo erectus*) emerged in Africa roughly two million years ago through a speciation event.[13] Throughout the Pleistocene, human populations across the Old World remained more or less connected by gene flow sufficient to prevent speciation. Populations were likely to have been more fragmented at times and more differentiated than living humans, but enough gene flow occurred to keep most, if not all, part of the same evolving species. Local populations frequently became extinct and were replaced by migrants from neighboring populations, thereby reducing the species effective population size.

I further hypothesize that for most of this time Africa remained the most heavily populated region in terms of both population size and density. This does not mean that relative population sizes were constant—I doubt that to be the case—but that *on average* Africa remained the core. Following Fred Smith, I suggest that the initial change to modernity occurred in Africa and then spread throughout the rest of the

Old World via gene flow,[14] although I suggest that the nature of gene flow is not clear. Was it due to the exchange of genes between local populations as envisioned by the isolation by distance model, or was it the dispersal of populations out of Africa into different regions followed by admixture with archaic humans outside of Africa? I suggest that *both* forms of genetic exchange occurred and their relative roles varied across both time and space. Gene flow may have been more regular within certain regions, such as Africa, but more sporadic in other areas. We need additional fossil and archaeological data to investigate this problem further, but the bottom line is a suggestion that several (perhaps all) regions remained part of a single species because of gene flow, be it regular or occasional.

The observed Late Pleistocene "population explosion" of 40,000–50,000 years ago can be explained in terms of changing population dynamics.[15] I suggest that the increase in species effective size was because of a combination of: 1. An actual increase in census size, 2. Increased gene flow (both among local groups and long-distance), and 3. An increase in the survival of local populations (i.e., a decrease in the rate of local extinction and recolonization). The latter two changes did not increase census size *per se* but instead increased the ratio of effective size to census size, causing a rate of increase in effective population size that was greater than the actual rate of increase in census population size. As species effective size increased, the balance between gene flow and genetic drift would change; genetic drift would be reduced, and continued gene flow would lead to greater homogeneity of populations (lowered F_{ST}), which in turn would contribute further to the growth of species effective size in a classic positive feedback loop. A summary of this process is shown in Fig. 9.2. It is tempting to suggest that this expansion took place first in Africa, based in part on some genetic evidence for an earlier African expansion and the evidence for bone tools (usually associated with modern humans in the Upper Paleolithic) appearing earlier in Africa. However, pending further genetic and archaeological support, this idea must be taken as suggestive only.

What started this process? Was the initial "trigger" the origin of, or continued evolution of, some biological and/or cultural adaptation(s)? Some have suggested that the success of anatomically modern humans rests in some basic adaptive biological change that began to take place over 100,000 years ago in Africa. Suggestions have ranged from reduced robusticity to shifts in regulatory genes to increased linguistic ability, but there has been no consensus. Others have suggested the Pleistocene population expansion essentially reflects cultural changes, primarily improvements in lithic and other technologies. There is also considerable debate about the speed of cultural changes—did they appear rapidly, or are we seeing the acceleration of earlier developments (e.g., bone tools in Africa dating back 90,000 years)? Given the association of early anatomically modern humans with Middle Paleolithic tools, and the close correspondence of the population explosion with the spread of Upper Paleolithic culture, it is tempting to suggest that the population explosion was initially triggered by cultural, rather than biological, causes. Given the wide confidence intervals on estimated expansion dates, it is wise to treat this as a hypothesis and not a conclusion.

As the effective population size of our species increased, the balance between gene flow and genetic drift changed (as well as the role of natural selection, although I am confining my remarks at present to patterns inferred from presumably neutral traits). Differentiation due to genetic drift began to decline in intensity, and

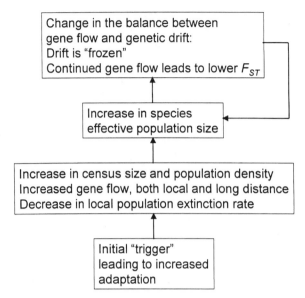

Figure 9.2. Flow chart modeling possible genetic effects of changing patterns of migration and population size during the Late Pleistocene.

increased gene flow began to counter the effects of previous genetic drift.[16] As a result, I suggest that populations of living humans are less differentiated than they would have been in the past. As this process took place, the ratio of effective size to census size would increase further, and the process would rapidly accelerate, acting to continue to reduce F_{ST} and make the entire system driven more and more by gene flow.

I suggest that increased species effective size and gene flow countered and reversed previous isolation and differentiation. There are many examples in evolutionary biology of isolation and genetic differentiation leading to new, and reproductively separate, species. Isolation and the formation of distinct subspecies are often seen as the first step in the speciation process. If my interpretations were correct, then why wouldn't isolation and differentiation of *Homo* during much of the Pleistocene lead to speciation? We see a number of examples of subspecific differentiation in other primates. A good example is the gorilla, comprised of three physical and genetically rather distinct subspecies limited to specific environments. Their continued isolation means that even if they are not genetically different species, their recent evolution is likely to be one of continued isolation, making them independent evolutionary entities for all practical purposes. Early humans may have been similarity isolated and differentiated at times. The difference is that our ancestors did not remain isolated. From the time of early *Homo erectus*, humans have been capable of long-distance movement, even if on an occasional basis, driven in part by our biological capabilities but also due to our cultural adaptations.[17] As shown in earlier chapters, it would not take a lot of gene flow over long intervals of time to maintain human populations within the same species. This gene flow may have been sporadic in nature, perhaps consisting of occasional dispersals out of Africa, or it may have been more regular at times. In either event, I propose that

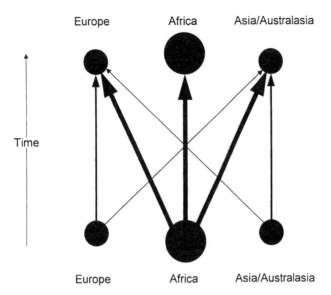

Figure 9.3. *The "Mostly Out of Africa" model, a multiregional model in which Africa contributes the most to accumulated ancestry in all regions. The width of the lines indicates possible relative contributions in terms of accumulated ancestry.*

sufficient levels of average gene flow kept the speciation process on hold until the rapid demographic changes in the Late Pleistocene eliminated this possibility and led to a relatively homogeneous species of living humans.[18]

I further suggest that, given the larger effective population size in Africa throughout this time, and especially after an expansion, continued gene flow would lead to greater total African ancestry in all regions. Our ancestry relative to 100,000–200,000 years ago may therefore be considered "mostly out of Africa," a model summarized in Fig. 9.3. To be fair, I should note that the African replacement model could be described in terms of the same processes, and the only difference is that none of the non-African ancestry of living humans can be traced back past 100,000 years ago. Again, I tend to view this as less probable given both some genetic evidence for ancient non-African ancestry and the fossil evidence for regional continuity.

If I am correct and a multiregional model is the probable answer, then *which* multiregional model is more probable—primary African origin or regional coalescence? By its very nature, a view of "mostly out of Africa" fits more closely with the models I have lumped under the category of "primary African origin" models because of its reliance on greater African genetic impact (because of size differences) and higher rates of gene flow out of Africa than into Africa. A model of regional coalescence is not necessarily ruled out, because there exists the possibility of some modern traits arising elsewhere.

The common element is that both primary African origin and regional coalescence models suggest some ancient genetic evidence outside of Africa in opposition of a model advocating speciation and complete replacement. Even if we could demonstrate only a very small contribution outside of Africa, the fundamental evolutionary process is still different from speciation and replacement. I suggest that our species can be described in terms of a "recent African origin," but within a multiregional framework.

9.4   ISSUES AND FUTURE DIRECTIONS

Much of what I have offered in this chapter are hypotheses that still need further testing and possible acceptance, modification, or rejection. Even the basics are not that clear; although I favor a multiregional interpretation as more probable, this does not mean that the African replacement model can be ruled out. To a large extent, my own interpretations are based on my reading of the fossil record and my own biases. For the most part, excepting some evidence of ancient non-African ancestry, the genetic data do not resolve the basic debate between speciation/replacement and multiregional evolution. I view the genetic data as more useful in telling us about demographic history, including changes in population size and migration patterns, than about resolving phylogenetic questions.

Apart from the continuing debate over modern human origin models, some basic questions need to be addressed. Some of these questions apply to both models, and others are contingent upon one or the other. A partial list includes the following:

- If I am wrong, and the African replacement mode is correct, then *why* did total replacement occur? Was it the consequence of a new biological adaptation, cultural adaptation, or both? If, as some evidence suggests, archaic and modern humans frequently used and shared the same technology, then what gave the moderns an edge? Was previous genetic differentiation sufficient to make them incapable of sharing genes (different species) even if they did come into contact with one another?

- If a multiregional model is correct, then does this mean that *all* major geographic regions contributed genetically to modern humans? Is it possible that the mode of evolution was multiregional for some non-African regions, such as Australasia, but replacement of another species in other regions, such as the Neandertals? If so, are we dealing with an evolutionary history more complex than implied in an either-or answer to the question of cladogenesis and anagenesis? If continued research on fossils and mitochondrial DNA shows that the Neandertals *were* likely to have been a separate species, then does this imply *total* replacement? No, because a general view of multiregional evolution suggests that genetic contributions come from at least one region outside of Africa and not necessarily all of them.

- What was the nature of gene flow in the past, whether we are talking about short-term evolution within a single new species or more ancient migration under a multiregional model? Where and when did gene flow between local populations restricted by isolation by distance dominate? Were there actual movements of entire populations leading to a pattern of admixture? If so, how frequent and in which direction? Marta Lahr and Robert Foley[19] have suggested a variant of the African replacement model in which a single origin is followed by multiple dispersals out of Africa at different times and along different geographic routes. One could imagine a multiregional equivalent differing only in setting the time of the initial origin back two million years ago with *Homo erectus*. It is possible that multiple dispersals extending back this far could act to maintain populations across the Old World in a single species.

- Is the metapopulation biology model of extinction and recolonization of local populations viable for discussing human evolution? Were rates of local extinction large enough relative to rates of gene flow to increase differentiation enough to maintain a low ratio of effective population size to census size while maintaining a single species?
- Can we improve the evidence for the Pleistocene population expansion? Is it the same for all genes or DNA sequences, or do different patterns emerge because of different levels of diversity existing before an expansion?
- What caused the population expansion? Was it primarily biological or cultural in nature?
- How do more recent demographic events, such as the agricultural revolution, affect our recent genetic history and possibly obscure earlier events?

It seems to me that the answers to the search for modern human origins are unlikely to exist in any one study, or any one type of data. The answers, if we can figure them out, lie in the synthesis of many different fields, including but not limited to paleontology, archaeology, molecular genetics, and population genetics. I make no claim to having all of the answers; rather, it should be obvious that although I favor certain interpretations, I do so in the sense of relative probabilities rather than definitive conclusions. Perhaps I am too cautious, but my experience suggests that the story of human evolution still holds many surprises for us.

I hope that I have at least provided you with a short historical review of how genetic data have been used and can be used. My primary messages have been the need to consider alternatives (compatibility does not necessarily equal proof) and the need for more interdisciplinary discussion across all fields concerned with human origins and evolution. I fully expect new data bearing on these questions to be published before this book gets into print, and I take my chances that what I have written won't be terribly out of date by the time you read this. If so, just keep in mind that is the nature of science—it is not a place for those that want nothing but complete and final answers. Still, that's what makes it fun!

Chapter Notes

CHAPTER 1: REFLECTIONS OF THE PAST

1. Some general introductory textbooks describing the field of biological anthropology (also known as physical anthropology) include Boyd and Silk (1999), Jurmain et al. (1999), Park (1998), Relethford (2000b), and Stein and Rowe (1999).
2. Harpending (1974, p. 229).
3. Cann et al. (1987).

CHAPTER 2: EVOLUTION AND GENETIC HISTORY

1. The fundamentals of basic molecular and Mendelian genetics can be found in the biological anthropology texts listed in the chapter notes for Chapter 1. Additional and more comprehensive information can be found in virtually any college-level genetics text, such as Russell (1996) and Strachan and Read (1999).
2. Basic information on DNA methods can be found in Brown (1999) and Strachan and Read (1999). Additional information on anthropological uses can be found in Williams (1989) and Devor (1992) for restriction fragment length polymorphisms (RFLPs), Stoneking (1993) for mitochondrial DNA, and Hammer and Zegura (1996) for Y-chromosome polymorphisms.
3. In addition to general texts in biological anthropology and genetics, more detailed mathematical explanations of population genetics and microevolution can be found in most college-level population genetics texts, including Crow and Kimura (1970), Nei (1987), and Hartl and Clark (1997), among others.
4. The local and regional sizes of hunting-gathering groups obviously varies from one case to the next, but average figures of 25 for the local band and 500 for a tribal unit are useful approximations for much of human evolutionary research (Birdsell 1968; Hassan 1981).
5. Detailed explanations of the macroevolutionary process, including its relationship to microevolution (often debated), can be found in any college-level evolution text, such as

Fututma (1997). There are a number of philosophical and scientific debates about definitions of species, ranging from typical application of the biological species concept to the use of the evolutionary species concept. This debate is beyond the scope of this text, and the edited volume by Ereshefsky (1992), which contains a number of useful essays on these topics, should be consulted.

6. Details of the original Harvard study in the 1930s are found in Hooton et al. (1955). The primary study conducted by Mike Crawford and myself is described in Relethford and Crawford (1995) with additional information in Relethford et al. (1997). This research was supported in part by research grant no. DBS-9120185 from the National Science Foundation.

7. See Williams-Blangero and Blangero (1989), Relethford and Blangero (1990), and Relethford et al. (1997) for details on the relationship between anthropometric variation and genetic distances.

8. See North et al. (2000) for an analysis of Irish history based on blood groups.

9. Ruvolo et al. (1993).

10. An excellent accounting of the early history of debate on the ape-human split, dealing with both the fossil and genetic data, is found in Lewin (1997). A recent review of the current status of the field can be found in Conroy (1997). Key original papers include Sarich and Wilson (1967a and 1967b).

11. Sarich (1971, p. 76).

12. I used the terms "population structure" and "phylogenetic branching" in a review article (Relethford 1998a) to distinguish between two approaches to using genetic data to reconstruct population history. The population structure approach looks at genetic drift and gene flow operating on a set of populations. The phylogenetic approach is used when populations evolve through a fissioning process, such as the formation of a new species.

CHAPTER 3: THE MODERN HUMAN ORIGINS DEBATE

1. Basic details on the fossil record of human evolution can be found in the introductory biological anthropology texts listed in Chapter 1. Additional detail can be found in several advanced college-level texts, including Conroy (1997), Poirier and McKee (1999), Klein (1999), Wolpoff (1999), and Campbell and Loy (2000). The archaeological record is covered in detail in Schick and Toth (1993) and Klein (1999).

2. The small proportion of earth's history taken up by human evolution is often difficult to grasp because of the enormity of the numbers involved. Even though we know intellectually that a billion years is 1000 million years, we often wind up lumping both "billion" and "million" together as simply very large numbers. The late Carl Sagan (1977) developed his "Cosmic Calendar" as a device to help put such large numbers into everyday perspective. He took the history of the universe and compressed it into a yearly calendar. Another method that I find useful is compressing earth's history into a 100-yard football field. We let one end of the field represent 4.6 billion years ago and the other end today. On this time scale, you would have to walk 88 yards to reach the beginnings of the Paleozoic Era 545 million years ago (mya), and 95 yards to reach the beginning of the Mesozoic Era 245 mya. The Cenozoic Era began 65 mya and would represent 98.5 yards on this time scale. The first hominids (4.2–4.4 mya) occur at 99 yards, 2 feet, and 9 inches. The first stone tools (2.5 mya) would be at 99 yards, 2 feet, and 10 inches. The debate over modern human origins focuses a lot on events during the last 200,000 years; this seems like a long time to us, but on this scale it would cover roughly only the last quarter-inch of the football field.

3. In the past, traditional textbook explanations presented a simplified view where *Homo habilis* evolved into *Homo erectus*. The realization of greater than expected variation among early *Homo* has led many anthropologists to reclassify some of the *H. habilis* specimens as *Homo rudolfensis*. As a result, it is not clear which of these species (if either) is the best choice for classifying the earliest fragmentary remains of the genus *Homo*. At the same time, the fossil record for *Homo erectus* has been pushed back to close to two million years (Wolpoff 1999). Our current reading is for the existence of three species of *Homo* by two million years ago in Africa, of which the species *Homo erectus* is the most likely ancestor of later humans. In fact, recent analyses and reviews by Wolpoff (1999) and Wood and Collard (1999) suggest reclassifying *H. habilis* and *H. rudolfensis* into the genus *Australopithecus* (in which case the two species would then be known as *Australopithecus habilis* and *Australopithecus rudolfensis*).

4. More detailed information regarding *Homo erectus* can be found in Rightmire (1990, 1992). A nicely written summary of recent discoveries and analyses, focusing on the complete KNM-WT 15000 skeleton from Kenya, can be found in Walker and Shipman (1996).

5. Swisher et al. (1994).

6. Summarized in Balter and Gibbons (2000).

7. Gabunia et al. (2000).

8. Tattersall (1995); Wood and Collard (1999).

9. Rightmire (1990, 1992); Kramer (1993).

10. Hou et al. (2000) provide recent evidence for the presence of Acheulian-like tools in China.

11. The new dating estimates for the Ngandong hominids are reported in Swisher et al. (1996). See also Grün and Thorne (1997) and Swisher et al. (1997) for further discussion and debate over the accuracy of these dates.

12. Figure 3.5 plots cranial capacity (in centimeters) as a function of time in 136 *Homo* specimens reported by Ruff et al. (1997). These specimens range in age from 1000 years ago to 1.8 million years ago. The solid line in the figure is the best fitting curve obtained using a method known as piecewise nonlinear regression. In mathematical terms, this line is described by the equation

$$y = a - b(x - c)(x < c)$$

where y is cranial capacity (cm^3), x is the geological date (in thousands of years), and a, b, and c are parameters estimated from the regression analysis. The relationship $(x < c)$ is equal to 1 if $x < c$ and equal to 0 if $x \geq c$. The parameter c corresponds to the date at which the curves change. This means that the regression equation is equal to a constant value of

$$y = a$$

when $x \geq c$ (which means that the geologic date is older than c) and increases over time according to the equation

$$y = a' - bx$$

when $x < c$ (which means that the geologic date is younger than c) and where

$$a' = a + bc$$

The regression equation was fit using the nonlinear regression method in the SYSTAT computer program (Version 8, SPSS, Inc.). This resulted in the following parameter estimates: $a = 882$ (±46.4), $b = 0.79$ (±0.11), and $c = 700$ (±93.5). The cutoff point (c) is the estimated time at which the sharp increase in cranial capacity began, in thousands of years. These parameters mean that cranial capacity of *Homo* changed little from an average of 882 cm^3 until 700,000 years ago, at which time it began increasing.

13. Additional background on the anatomy and culture of the Neandertals can be found in Trinkaus and Shipman (1992), Stringer and Gamble (1993), and Shreeve (1995), in addition to the sources on human evolution already cited. Note that there are two spellings— "Neandertal" and "Neander*th*al"—although the *h* is silent and both are pronounced the same way.

14. The most recent date for Neandertal fossils is now 28,000 years ago, as reported by Smith et al. (1999).

15. There is debate over the definitions of "archaic" and "modern" human anatomy, with some arguing for a clear distinction (e.g., Stringer 1985, 1994; Stringer and Gamble 1993) and others arguing that modernity is a concept that can be applied to specific *traits* but not specific *populations* (Wolpoff and Caspari 1997). Willermet and Hill (1997) suggest that clear-cut distinctions are difficult between evolutionarily related groups because the definitions rely on relative changes (e.g., *reduced* size in brow ridges) rather than presence or absence. They further suggest application of "fuzzy logic," which does not require categories to be absolutely distinguished. Although preliminary, their paper suggests that further incorporation of fuzzy logic in paleoanthropology has much potential.

16. A number of edited volumes that have been published in the past two decades present detailed arguments regarding interpretations of the fossil and archaeological evidence for the emergence of modern humans, including discussions of anatomic status and geologic dating. These volumes are useful for reference because they present alternative views on the same data. Some examples include Smith and Spencer (1984), Mellars and Stringer (1989), Bräuer and Smith (1992), Aitken et al. (1993), Nitecki and Nitecki (1994), Clark and Willermet (1997), and Omoto and Tobias (1998).

17. A lively discussion on the nature of species and the process of human macroevolution, representing two different viewpoints, can be found in the exchanges between Tattersall (1994a, 1994b) and Wolpoff (1994a, 1994b). See also Tattersall (1986, 1992).

18. Wolpoff et al. (1993) argues that, instead of using the species names *Homo erectus* and *Homo sapiens* as labels for stages in an evolving lineage, we should use the evolutionary species concept and label the *entire* lineage as *Homo sapiens*.

19. King (1864), reprinted in Meikle and Parker (1994).

20. A thorough history of changing views on the evolutionary status of Neandertals is found in Trinkaus and Shipman (1992).

21. Many papers have been written on the basic premises of modern human origin models, such that a complete description and analysis would take an entire volume. One of the earliest papers to identify the notion of a recent African origin and contrast it with a model of gene flow among populations across the Old World was that of Howells (1976), who termed the "Noah's Ark" hypothesis as a model in which modern humans appeared in one place and then expanded across the Old World. One of the later, and very influential, papers advocating African replacement was that by Cann et al. (1987), based on the evidence from mitochondrial DNA. Shortly thereafter, the paper by Stringer and Andrews (1988) reviewed both genetic and fossil evidence to argue for a view of recent African evolution, suggesting little, if any, admixture of moderns emerging in Africa with preexisting archaic human populations outside of Africa. The multiregional evolution model was first developed by Wolpoff et al. (1984) as an explanation of evolutionary *process* describing worldwide evolution that allows both similar evolutionary trends (e.g., increase in cranial capacity) as well as regionally specific variants.

22. Many additional papers have since debated both the meaning and interpretations of modern human origins models, including Smith et al. (1989), Stoneking and Cann (1989), Wolpoff (1989), Stringer (1990, 1994), Thorne and Wolpoff (1992), Aiello (1993), Frayer et al. (1993, 1994), Simmons (1994), Stringer and Bräuer (1994), Wolpoff et al. (1994), Relethford (1995, 1998a, 1999, 2001b), Stringer and McKie (1996), Bräuer and Stringer (1997), Clark (1997), Smith and Harrold (1997), and Wolpoff and Caspari (1997), among others.

23. The issue of species differences in relationship to cladogenetic speciation is a further source of confusion since not everyone uses species names in the same way. Among evolutionary biologists of the cladistic school, species are reproductively unique entities that evolve via cladogenesis. Some geneticists and anthropologists (e.g., Cann et al. 1987; Harpending et al. 1998) consider the origin of modern humans a true speciation event, whereas for others it is less clear. Stringer, for example, has frequently suggested that modern humans represent a different species from archaic humans, and that the former replaced the latter with little, if any, admixture. However, Bräuer and Stringer (1997) state that their recent African origin model is not the same as speciation and replacement, noting that "this again is far from reality. For example, whether we regard the morphological differences between the Neanderthals and modern humans as intraspecific (GB) or interspecific (CBS), we are certainly not dealing with a biospecies concept with absolute fertility boundaries, but only with paleospecies." (p. 193). Elsewhere, they also note that "Neither of us feels that taking a RAO position necessarily entails accepting that dispersing modern humans *must* have represented a new species." (pages 416–417, italics in original). This view stands apart from the total replacement view advocated by Cann et al. (1987) and others that suggest that total replacement *did* happen.

24. Zubrow (1989).

25. The classic original paper on multiregional evolution is Wolpoff et al. (1984), which was written before many of the newer arguments suggesting the evolution of modern humans through speciation and total replacement. Consequently, the initial development of multiregional evolution was concerned more with a *general* process. The general multiregional view has evolved somewhat over the years and can actually encompass a variety of different scenarios (see Wolpoff et al. 1994 and Wolpoff and Caspari 1997). I suspect that part of the debate arises from confusion over the general model (focusing on evolution *within* a single species) and the more specific view of regional coalescence expressed in some papers.

26. The multiregional model has been badly misinterpreted over the years. One major misconception is that regional continuity necessarily implies that the *majority* of ancestors for a region be from the same region. Taken to an extreme, this misconception leads to the further misconception that multiregional evolution is the independent and parallel evolution of humans in different regions. Neither of these misconceptions is correct (Wolpoff et al. 2000).

27. Some good data sources on genetic variation in living human populations include Roychoudhury and Nei (1988) and Cavalli-Sforza et al. (1994).

28. A recent statistical review of the geographic distribution of human skin color is Relethford (1997a).

29. Lahr (1994).

30. My representation of the models of modern human origins (Fig. 3.14) shows how I currently view the debate, although I do not expect everyone to agree with the specific way in which I have framed the models. This is my attempt to try to resolve some of the confusion over the way in which different models have been proposed and compared and the conflict in the way in which different terms have been used. Many previous works have focused on a comparison between recent African origin (RAO) and multiregional

evolution (MRE). According to my interpretation, RAO refers to one possibility for the location and timing of the archaic-modern transition, whereas MRE refers to one possibility of the mode of the transition. African replacement is by definition a RAO model, although not all RAO models are necessarily replacement. Likewise, some MRE models can also be RAO. I suggest using RAO to refer specifically to a recent African origin of a modern *population* and MRE to refer specifically to any model that considers the transition within a single species.

31. My "African replacement" model is equivalent to the views of Cann et al. (1987) and was labeled by Frayer et al. (1993, 1994) as "Eve theory." Stringer's (Stringer 1990, 1994; Stringer and Andrews 1988; Stringer and McKie 1996) views generally fit here, although he notes that his views do not necessitate a speciation event in the sense I am using it here (Stringer and Bräuer 1994, Bräuer and Stringer 1997). I would include Bräuer's hybridization model and Smith's assimilation model under what I call "Primary African origin" because they both argue for the emergence of modern human *populations* within an evolutionary lineage. Although Wolpoff has argued frequently for a specific interpretation of regional coalescence (e.g., Wolpoff and Caspari 1997; Wolpoff 1999), it is also clear that his more *general* view of multiregional evolution can also encompass primary African origin, although he (and others) doesn't necessarily think this is the correct interpretation (see Wolpoff et al. 1994).

32. The quotation about different views of multiregional evolution is from Wolpoff et al. (1994, p. 178). Italics are in the original. FHS refers to one of the authors (Fred H. Smith).

33. Smith et al. (1989).

34. Bräuer (1984, 1992).

35. The exchange in *American Anthropologist* is found in Frayer et al. (1993, 1994) and Stringer and Bräuer (1994). See also Bräuer and Stringer (1997).

CHAPTER 4: IN SEARCH OF OUR COMMON ANCESTOR

1. Vigilant et al. (1991).

2. Cann et al. (1987).

3. Cann et al. (1987, p. 35–36).

4. Wainscoat (1987).

5. Tierney et al. (1988).

6. Brown (1990).

7. Stringer and Andrews (1988).

8. Wolpoff (1989).

9. See Vigilant et al. (1991).

10. The problems with the tree reconstructions of Cann et al. and Vigilant et al. were pointed out by Maddison (1991), Templeton (1992), and Hedges et al. (1992).

11. Hedges et al. (1992, p. 739).

12. Penny et al. (1995).

13. Gagneux (1999).

14. Gagneux (1999, p. 5080).

15. Templeton's critique is Templeton (1993); see also Stoneking's (1994) response and Templeton's (1994) rebuttal. See also Excoffier and Langaney (1989) for another examination of the geographic location of "Eve." See Templeton (1997a, 1997b, 1998) for further information on his geographic analysis.

16. Stoneking et al. (1992) used a different method of tree construction than in the Cann et al. (1987) study and calibrated it using the known date of the colonization of New Guinea.

17. Horai et al. (1995).

18. Wills (1995).

19. An excellent review of coalescent theory is Hudson (1990). Other useful papers include Donnelly and Tavaré (1995), Donnelly (1996), Harding (1996, 1997), and Marjoram and Donnely (1997).

20. Templeton (1993).

21. Ruvolo et al. (1993).

22. Some have suggested evidence for selection on part of the mitochondrial DNA genome (e.g., Templeton 1996; Hey 1997; Loewe and Scherer 1997; Wise et al. 1997; Hawks, Hunley et al. 2000).

23. Awadalla et al. (1999).

24. Kivisild and Villems (2000); Jorde and Bamshad (2000); Kumar et al. (2000); Parsons and Irwin (2000). See also the response by Awadalla et al. (2000).

25. Dorit et al. (1995).

26. Hammer (1995).

27. Hammer et al. (1998).

28. Underhill et al. (1997).

29. Thomson et al. (2000).

30. Huang et al. (1998).

31. Kaessmann et al. (1999).

32. Harris and Hey (1999).

33. Harding et al. (1997).

34. See Hawks, Hunley et al. (2000) for a review of other nuclear DNA coalescent dates.

35. Rogers and Jorde (1995).

36. Ruvolo (1996).

37. Hammer et al. (1998).

38. Harris and Hey (1999).

39. Harding et al. (1997).

40. Harding et al. (1997, p. 782).

41. Insman et al. (2000).

CHAPTER 5: GENETIC DIVERSITY AND RECENT HUMAN EVOLUTION

1. O'Brien et al. (1983).

2. The classic example of selection for the heterozygote in humans is the case of the sickle cell allele for the beta protein chain of hemoglobin. The normal form of the gene is the *A* allele. There are several known mutations of this gene, including the sickle cell gene (*S*), which results from a single point mutation. Individuals with the *SS* genotype are likely to die from sickle cell anemia, thus tending to eliminate the *S* allele from a population. In human populations native to regions where malaria is epidemic, the frequency of *S* is higher, due to the fact that the heterozygotes (genotype *AS*) have an advantage by having their biochemistry altered enough to make them less susceptible to malaria. Individuals with the genotype *AA* are more likely to suffer from malaria, whereas individuals with the genotype *SS* have sickle cell anemia. The heterozygotes (*AS*) have the

highest fitness in such environments, and evolution has led to a balance in the frequencies of A and S such that survival is maximized.

3. Figure 5.1 shows the evolution of heterozygosity (a measure of diversity discussed later in Chapter 5) over time. The values were obtained using an iterative equation that expresses heterozygosity in a given generation (H') as a function of heterozygosity in the previous generation, the rate of gene flow (m), the mutation rate (μ), and the number of reproductive adults (N):

$$H' = 1 - (1 - m - \mu)^2[1 - H(1 - 1/2N)]$$

(see Hartl and Clark 1997). For the example in Fig. 5.1, $m = 0.001$, $\mu = 0.0001$, and $N = 100$.

4. The expected amount of heterozygosity for a locus with two alleles [$H = 2p(1 - p)$] is a consequence of the mathematical relationship known as Hardy–Weinberg equilibrium, discussed in detail in any population genetics text. If there are two alleles (call them A and a) with frequencies p and q, the probability that someone will have the genotype Aa is easy to compute. There are two ways to have this genotype—you could inherit an A allele from your mother and an a allele from your father, or you could inherit an A allele from your father and an a allele from your mother. If the allele frequencies are the same in both sexes, then the probability of receiving an A allele is by definition equal to p and the probability of receiving an a allele is by definition equal to q. Because there are only two alleles, the sum of the allele frequencies is equal to $p + q = 1$, which means that the frequency of inheriting an a allele can also be expressed as $q = 1 - p$. The probability of inheriting an A allele from your mother *and* an a allele from your father is the product $pq = p(1 - p)$. Likewise, the probability of inheriting an A allele from your father *and* an a allele from your mother is also equal to $pq = p(1 - p)$. Now, the probability of having two different alleles (A from one parent and a from the other) is equal to the probability of either event occurring, which is

$$p(1 - 1p) + p(1 - p) = 2p(1 - p)$$

5. The allele frequencies for the alpha haptoglobin gene for Nigeria were taken from Roychoudhury and Nei (1988).

6. The formal equation for heterozygosity (H) is

$$H = 1 - \frac{\sum p^2}{n}$$

where summation (Σ) is over all alleles for all loci, and n is the number of loci.

7. See Nei (1987) for a review of different measures of diversity for DNA sequences and other genetic polymorphisms.

8. Regional estimates of diversity are obtained in one of two ways—pooling all individuals within a region and averaging diversity across local populations within a region. The latter is preferred in some contexts because the former could possibly be affected by variation among populations within regions. In addition, some populations rely on a single population to represent a given region, which could be in error if that group is not truly representative of the region to which it belongs. Despite these problems, the general pattern for mtDNA, microsatellite DNA, craniometrics, and skin color is that there appears to be more genetic diversity within sub-Saharan Africa than in other geographic regions. Future studies must expand the data base and more closely investigate intraregional patterns of diversity.

9. Higher African mitochondrial DNA diversity has been reported by a number of studies, including those of Cann et al. (1987), Vigilant et al. (1991), Bowcock et al. (1994), and Jorde et al. (1995). The statistical significance of some of the higher African diversity has been questioned by Templeton (1993).

10. Higher African diversity for microsatellite DNA has been reported by a number of studies, including those of Bowcock et al. (1994), Deka et al. (1995), Jorde et al. (1995, 1997), Tishkoff et al. (1996), Pérez-Lezaun et al. (1997), and Relethford and Jorde (1999). The study by Deka et al. (1995) listed heterozygosities by local populations. In most cases, each region was represented by only one group, but in two cases (Europe, New World) two samples were listed. The values reported in Fig. 5.3 are based on the averages in those cases.

11. Higher African diversity has been found in two studies of quantitative traits: craniometrics (Relethford and Harpending 1994) and skin color (Relethford 2000a).

12. Stoneking et al. (1997).

13. Rogers and Jorde (1996).

14. Relethford (1997b).

15. Tishkoff et al. (1996).

16. See the news article in the same issue as the Tishkoff article (Fischman 1996) for reactions and alternative explanations.

17. Stoneking and Cann (1989, p. 22).

18. In the formula used to model mitochondrial DNA diversity, the number of pairwise nucleotide site differences in generation t, $m(t)$, as a function of time is

$$m(t) = \theta + [m(0) - \theta]e^{-t/N_f}$$

where $m(0)$ is the initial mtDNA diversity in generation $t = 0$, N_f is the number of reproductive females in the population, t is the number of generations, and θ is the level of diversity at equilibrium, which is

$$\theta = 2N_f\mu$$

and where μ is the aggregate mutation rate over all nucleotide sites. Figure 5.6 uses a rate of $\mu = 0.0015$ and sets the initial level of diversity $m(0)$ equal to zero for both populations. See Li (1977) and Rogers and Jorde (1995) for details on this model.

19. Figure 5.7 was based on the model above but modified to simulate an initial bottleneck by setting $m(0)$ equal to the diversity expected before the bottleneck ($N_f = 5000$), assuming this population had already reached equilibrium, and the new equilibrium value θ equal to the diversity expected with the new population size ($N_f = 50$). Using a mutation rate of $\mu = 0.0015$, this means that

$$m(0) = 2(5000)(0.0015) = 15$$
$$\theta = 2(50)(0.0015) = 0.15$$

which when inserted into the standard model gives

$$m(t) = \theta + [m(0) - \theta]e^{-t/N_f} = 0.15 + 14.85e^{-t/50}$$

where t is the number of generations after the bottleneck. The same approach was used for Fig. 5.8 for different values of female population size (and postbottleneck equilibrium diversity).

20. Figures 5.9 and 5.10 were constructed using the standard model but altering the population size and new equilibrium values in the equation when the population increased in size.

21. Models of sudden or instantaneous expansion in population size are mathematically convenient but unrealistic for any real species. Why bother with the math then? As is often the case, the simple (and unrealistic) mathematical model provides a reasonable approximation to reality. In this case, the simple model gives results similar to more complex and realistic, though mathematically cumbersome, models. See Rogers and Harpending (1992) for a related discussion of sudden expansion models.

22. As shown by Rogers and Jorde (1995), the half-life to convergence is the time in generations needed to reduce diversity half of the way from the current value $m(0)$ to a new equilibrium θ. This occurs when

$$\frac{m(t)-\theta}{m(0)-\theta} = 0.5$$

Plugging in the standard formula for $m(t)$ gives

$$e^{-t/N_f} = 0.5$$

Taking the natural logarithm of both sides gives

$$t = -N_f \ln(0.5) = 0.693 N_f$$

The implications of diversity reduction are discussed in Rogers and Jorde (1995) and in Relethford and Jorde (1999).

23. Population Reference Bureau web site
    (http://www.prb.org/pubs/wpds99/wpds99_1.htm).

24. Regional estimates of usable land mass during the Upper Paleolithic were developed by Birdsell as discussed by Hassan (1981).

25. Hassan (1981).

26. Thorne et al. (1993) also discusses the case for larger population size in sub-Saharan Africa.

27. My work on estimating relative regional population sizes is described in Relethford and Harpending (1995) using an extension of a model first developed by Harpending and Ward (1982). The Harpending–Ward model predicts that the expected average per-locus heterozygosity in population i is

$$E[\overline{H}_i] = \overline{H}_T(1 - r_{ii})$$

where $\overline{H}_T$ is the observed total heterozygosity obtained using the average allele frequencies pooled over all populations ($\overline{p}$) and r_{ii} is the genetic distance of population i to the pooled average allele frequencies

$$r_{ii} = \frac{(p_i - \overline{p})^2}{\overline{p}(1 - \overline{p})}$$

This value is computed for each allele, and then a final value is averaged over all alleles. Both total heterozygosity and the distance to the centroid involve the average allele frequencies, $\bar{p}$, which are weighted by population size. For any given allele, the mean allele frequency is

$$\bar{p} = \sum_i w_i p_i$$

summation is over all populations in the analysis, and w_i is the relative population size of population i, computed as

$$w_i = N_i \Big/ \sum_i N_i$$

where N_i is the population size of population i. The *expected* heterozygosity of population i can be compared to the *observed* heterozygosity for that population. Harpending and Ward (1982) show that the two values (observed and expected) should be the same *if* each population receives the same amount of gene flow from outside the area of analysis. Comparing observed and expected heterozygosities therefore provides us with a way of detecting differences in external gene flow. Relethford and Blangero (1990) extended the Harpending–Ward method for use with quantitative traits.

28. Relethford and Harpending (1994).
29. Relethford and Jorde (1999).
30. Bar-Yosef (1994).
31. Klein (1999).
32. Wolpoff and Caspari (1997, p. 306).

CHAPTER 6: GENETIC DIFFERENCES BETWEEN HUMAN POPULATIONS

1. See Harpending and Jenkins (1973) for details on their genetic distance measure and related statistics (see Workman et al. 1973 for additional details). The genetic distance between populations is related to the genetic distance of each population to the centroid (r_{ii}) discussed in Chapter 5, which in turn is related to an **R** matrix, which expresses the average relative genetic similarity between pairs of populations. The elements of the **R** matrix are the genetic similarities between populations i and j for a given allele, computed as

$$r_{ij} = \frac{(p_i - \bar{p})(p_j - \bar{p})}{\bar{p}(1 - \bar{p})}$$

where p_i and p_j are the allele frequencies for populations i and j, respectively, and $\bar{p}$ is the weighted average allele frequency as defined in Chapter 5. The elements of the overall **R** matrix are then obtained by averaging the above values over all alleles. The **R** matrix provides a measure of genetic similarity relative to the mean allele frequencies. When a pair of populations each have allele frequencies equal to the average, the corresponding element of the **R** matrix is zero. Positive elements of the **R** matrix indicate a pair of populations more closely related to each other than on average, and negative elements of the **R** matrix indicate a pair of populations less closely related to each other than on average. An **R** matrix (showing genetic similarity) is easily transformed into a distance matrix, usually denoted as $\mathbf{D}^2$ (showing genetic *dissimilarity*), using the formula

$$d_{ij}^2 = r_{ii} + r_{jj} - 2r_{ij}$$

A little algebra shows that, for any given allele, the genetic distance between populations i and j is equal to the value given in the text

$$
\begin{aligned}
d_{ij}^2 &= r_{ii} + r_{jj} - 2r_{ij} \\
&= \frac{(p_i - \overline{p})(p_i - \overline{p})}{\overline{p}(1-\overline{p})} + \frac{(p_j - \overline{p})(p_j - \overline{p})}{\overline{p}(1-\overline{p})} - 2\frac{(p_i - \overline{p})(p_j - \overline{p})}{\overline{p}(1-\overline{p})} \\
&= \frac{(p_i - p_j)^2}{\overline{p}(1-\overline{p})}
\end{aligned}
$$

2. F_{ST} was originally developed by Sewall Wright (see, e.g., Wright 1951).

3. F_{ST} can also be expressed in terms of the **R** matrix, as the weighted average diagonal value ($i = j$), such that

$$F_{ST} = \sum_i w_i r_{ii}$$

where summation is over all groups. F_{ST} thus measures the average genetic distance to the centroid.

4. The formula for the evolution of F_{ST} in Fig. 6.4 is an iterative equation expressing the value of F_{ST} in the next generation (F_{ST}') as a function of the current value of F_{ST}

$$F_{ST}' = (1 - m - \mu)^2 \left[F_{ST} + \frac{1 - F_{ST}}{2N} \right]$$

where N is the population size, m is the rate of gene flow per generation, and μ is the mutation rate (see Hartl and Clark 1997). This formula applies to diploid traits; different formulae are needed for mtDNA. In this example, $N = 100$, $m = 0.01$, and $\mu = 0.00001$. F_{ST} is initially set equal to zero in the starting generation. F_{ST} increases over time because of the influence of genetic drift but is also countered by the effect of gene flow and mutation, until an equilibrium is reached such that the loss of diversity is offset by the accumulation of diversity each generation. This model assumes an infinite number of populations and is modified for more realistic situations with a small number of groups, as shown by Rogers and Harpending (1986). Also see Harpending et al. (1996) for extensions to haploid traits and DNA sequences.

5. The first study to estimate F_{ST} from classical genetic markers was that of Lewontin (1972), who found a value of 0.06. Later studies using more complete data tend to show values closer to 0.10–0.12. F_{ST} has been estimated for major geographic regions (usually ranging from 3 to 6 regions) from classical genetic markers (Latter 1980; Nei and Roychoudhury 1982; Ryman et al. 1983; Livshits and Nei 1990; Cavalli-Sforza et al. 1994), RFLP loci (Bowcock et al. 1991; Jorde et al. 1995; Barbujani et al. 1997), microsatellite DNA (Deka et al. 1995; Barbujani et al. 1997), *Alu* insertion polymorphisms (Batzer et al. 1994; Stoneking et al. 1997), and craniometrics (Relethford 1994). Some analyses of microsatellite DNA (Jorde et al. 1995; Pérez-Lezaun et al. 1997) and mitochondrial DNA (Whittam et al. 1986; Jorde et al. 1995; Harpending et al. 1996) have found lower estimates of F_{ST} that may reflect higher mutation rates for these loci. Some of these papers also look at a more elaborate partitioning of genetic variation, by looking at variation among regions, among local populations within regions, and within local populations. Barbujani et al.

(1997) reviewed some of these studies and found consistent results—about 10% of genetic variation occurs between major geographic regions, about 5% among local populations within regions, and the remainder (85%) within local populations. Some reviews (see, e.g., Jorde et al. 1998; Relethford 1998a) suggest higher values of F_{ST}, ranging up to 0.15, but these higher numbers are based on incorrectly adding the among-region and among-local population-within region variance components. A figure of $F_{ST} \approx 0.10$ to 0.12 is most appropriate for describing the proportion of variation *among* geographic regions.

6. Templeton (1998).

7. Relethford (1994).

8. The formula for estimating migrant numbers from F_{ST} derives from the work of Rogers and Harpending (1986), which allows prediction of F_{ST} from migration rates and population sizes. Assume g populations each of size n. Each population receives a fraction m of its genes from all other populations and a fraction $1 - m$ from itself. This means that the number of migrants into a population (from all other sources) is $M = nm$. Rogers and Harpending show that F_{ST} is

$$F_{ST} = \frac{1}{1 + \dfrac{4ngm_2}{g-1}}$$

where m_e, the effective migration rate, is

$$m_e = \frac{1 - \left(1 - \dfrac{gm}{g-1}\right)^2}{2}$$

when the rates of migration are the same between all pairs of groups and there is no mutation (Rogers and Jorde 1995). Given relatively low rates of m, the effective migration rate is approximately

$$m_e \approx \frac{gm}{g-1}$$

Using this approximation, and noting that $M = nm$, F_{ST} is approximately

$$F_{ST} \approx \frac{1}{1 + \dfrac{4g^2 M}{(g-1)^2}}$$

Solving for M gives

$$M \approx \frac{(1 - F_{ST})(g-1)^2}{4g^2 F_{ST}}$$

To further figure the actual number of migrants between each pair of population, divide M by $g - 1$.

9. Livshits and Nei (1990).

10. Relethford and Harpending (1994, 1995).

11. Many studies have shown sub-Saharan African populations to be more genetically divergent. These studies include analyses of classical genetic markers (e.g., Nei 1978; Nei and Livshits 1989; Cavalli-Sforza et al. 1994; Relethford and Harpending 1995), craniometrics (e.g., Lynch 1989; Relethford and Harpending 1994, 1995), RFLPs (e.g., Bowcock et al. 1991), microsatellite DNA (e.g., Bowcock et al. 1994; Deka et al. 1995; Jorde et al. 1995; Tishkoff et al. 1996), mitochondrial DNA (e.g., Vigilant et al. 1991; Jorde et al. 1995), Y chromosomes (e.g., Underhill et al. 1997), and *Alu* insertion polymorphisms (e.g., Batzer et al. 1994; Stoneking et al. 1997).

12. There are many different ways to construct a dendrogram from a matrix of genetic distances. The method used in this chapter is UPGMA, which stands for unweighted pair-group method using arithmetic averages. See Romesburg (1984) for a clear explanation of this and other clustering methods.

13. Imaizumi et al. (1973).

14. The problems of using genetic distances to estimate divergence dates for populations *within* a species are noted by Weiss and Maryuma (1976) and Weiss (1988) among others.

15. The problems and dangers of interpreting "tree" diagrams in terms of an underlying phylogenetic branching model has been noted in recent years by Relethford and Harpending (1994), Relethford (1995, 1998a), and Sherry and Batzer (1997) among others. Earlier work also noted this problem; a particularly succinct critique was made over 25 years ago by Henry Harpending, who noted that a "disorder in anthropology is Linnaeus envy represented in an earlier generation's fondness for resolving populations into a mixture of 'pure' races and in the generation's attraction to tree structures to represent relations among open hybridizing evolutionary units as phylogenetic relations among closed independent units." (1974, p. 238).

16. Figure 6.8 was constructed using genetic distances based on anthropometric data from five populations along the west coast of Ireland, as described by Relethford (1991b).

17. Felsenstein (1982, pp. 9–10).

18. Migration matrix models allow the prediction of genetic distance reflecting the balance between gene flow, genetic drift, and mutation. Several related methods exist (see, e.g., Bodmer and Cavalli-Sforza 1968; Smith 1969; Imaizumi et al. 1970; Morton 1973; Rogers and Harpending 1986). I use the Rogers–Harpending method here because it focuses on variation in allele frequencies relative to the contemporary gene pool and not variation relative to ancestral allele frequencies, which are typically not known. The Rogers-Harpending method predicts an **R** matrix from which genetic distances between populations can be derived as outlined earlier.

19. Principal coordinates analysis provides a two (or more)-dimensional representation of a matrix of genetic distances (or some other measure of similarity or dissimilarity). See Gower (1966) for computational details.

20. Cavalli-Sforza et al. (1994).

21. Eller (1999).

22. Heterozygosity values are related to the **R** matrix:

$$H_i = H_T (1 - r_{ii})$$

where H_i is the expected heterozygosity in population i, H_T is the total heterozygosity if all populations are pooled, and r_{ii} is the ith diagonal of the **R** matrix (Harpending and Ward 1982). Because the value of H_T cannot be obtained from migration matrix analysis, I use *relative* heterozygosity in Figs. 6.12–6.14, expressing the heterozygosity in population i relative to the total heterozygosity as

$$\frac{H_i}{H_T} = \frac{H_T(1-r_{ii})}{H_T} = (1-r_{ii})$$

The important thing to remember is that absolute and relative heterozygosity can be interpreted the same way—the larger the value, the more genetic diversity.

23. See Harpending et al. (1993) for a description of the "weak Garden of Eden" model. This model postulates an expansion in population size occurring *after* the initial origin and dispersal of *Homo sapiens* in Africa. By contrast, the "strong Garden of Eden" model suggests that regional divergence and population growth occurred at roughly the same time.

24. See Relethford and Harpending (1994) for the method used to estimate migration matrices from genetic distances and application of this method to craniometric data. See Relethford and Harpending (1995) for application of the method to classical genetic marker data.

25. The migrant numbers in Fig. 6.15 reproduce the pooled genetic distance matrix exactly. We further found that a simpler model, consisting of 0.33 migrants per generation between *each* pair of populations, provided an excellent fit to the distances (Relethford and Harpending 1995).

26. The model used here assumes symmetry in the *number* of migrants. This is a common and useful assumption when using migration matrices. If the *numbers* of migrants were not symmetric, then population sizes would change over time because there would be more migrants into a given population than out of it (Morton 1973). In actual application, imposing symmetry does not skew results very much (Rogers and Harpending 1986).

27. See Bowcock et al. (1991) for an interpretation that includes phylogenetic branching and admixture.

28. See Rogers and Harpending (1986) for derivation and discussion of the half-life statistic used in migration matrix analysis. See Rogers and Jorde (1995) for a more general discussion of half-life statistics in population genetics and their relevance to the modern human origins debate.

CHAPTER 7: HOW MANY ANCESTORS?

1. See Hassan (1981), Weiss (1984), and Livi-Bacci (1997) for estimates of the size of the human species in historic and prehistoric times.

2. See Harpending et al. (1993, 1998) and Rogers and Jorde (1995) for statements of the argument that small population size as estimated from genetic data supports an African replacement model. See Templeton (1997b, 1998), Relethford (1998a, 2001b), and Hawks, Hunley et al. (2000) for differing interpretations.

3. See Wobst (1974) and Jorde (1980) for data on the proportion of a population of reproductive age. Jorde in particular notes the variation that exists in human populations.

4. See Crow and Kimura (1970) and Hartl and Clark (1997) for detailed discussion about the concept and computation of effective population size.

5. Assuming no gene flow, the expected level of inbreeding in generation t is

$$F_t = F_{t-1} + \frac{F_{t-1}}{2N}$$

where F_{t-1} is the inbreeding in the previous generation and N is the breeding population size (Crow and Kimura 1970; Hartl and Clark 1997).

6. Hassan (1981).

7. Dates for the Lower, Middle, and Upper Paleolithic were taken from Feder (2000).

8. See Harpending et al. (1993) for derivation of their species census size, keeping in mind that their figure was designed to be a bare *minimum*. Also see Howell (1996) for a review of other estimates of ancient species size.

9. Nei and Grauer (1984).

10. Takahata (1993).

11. Templeton (1993, 1997b).

12. Sherry et al. (1997).

13. Ayala (1995).

14. Harding et al. (1997).

15. See Hawks, Hunley et al. (2000) for a review of estimates of species effective size.

16. Templeton (1997b, 1998).

17. These simulations were performed using Stephen Wooding's "TreeToy" Java applet available at http://www.anthro.utah.edu/popgen/programs/TreeToy.

18. Harpending et al. (1996, 1998).

19. The data for these analyses are described by Harpending et al. (1998) and were kindly provided by Henry Harpending and Lynn Jorde.

20. Harpending et al. (1998).

21. Di Rienzo et al. (1998).

22. Reich and Goldstein (1998).

23. Kimmel et al. (1998).

24. Hawks, Hunley et al. (2000).

25. The Rogers–Harpending method of dating the timing of the expansion (in size) of the human species was introduced in Rogers and Harpending (1992). Additional details and modifications are given in Rogers (1992, 1995, 1997), Harpending (1994), and Rogers et al. (1996).

26. Rogers (1995).

27. Sherry et al. (1994).

28. Excoffier and Schneider (1999).

29. Another analysis suggesting that Africa expanded in size earlier than outside of Africa is given in Relethford (1998b).

30. Di Rienzo et al. (1998).

31. Reich and Goldstein (1998).

32. See Rogers (1995, 1997) for comments on the likely preexpansion population size.

33. See Harpending et al. (1996, 1998) for description of the hourglass and long-necked bottle hypotheses of ancient population growth.

34. Harpending et al. (1993, p. 495).

35. Clark (1997).

36. Yellen et al. (1995).

37. See Ambrose (1998) for details on the Toba hypothesis and Rampino and Self (1992) for more details on the Toba eruption.

38. See Harding et al. (1997) and Sherry et al. (1997) for very ancient estimates of effective species population size. See Harpending et al. (1998) for conclusions based on the long-necked bottle model.

39. See Hawks, Hunley et al. (2000) for arguments against a Pleistocene expansion and their suggestion that low species effective size is a genetic signature of the speciation that led to *Homo erectus*, and not modern *Homo sapiens*.

40. Wright (1940), reprinted in Wright and Provine (1986, p. 355).

41. See Hanski and Gilpin (1997) and Hanski (1998) for reviews of metapopulation biology.

42. Takahata (1994, p. 805). See also Takahata (1995).

43. The effect of extinction and recolonization of local populations on effective species size has been considered by Slatkin (1977), Maryuma and Kimura (1980), Takahata (1994, 1995), Barton and Whitlock (1997), Hedrick and Gilpin (1997), and Whitlock and Barton (1997), among others.

44. The Whitlock-Barton model expresses species effective size as a function of F_{ST}:

$$N_e = \frac{Nn(1-\lambda)}{1 - F_{ST} + 2F_{ST}N\lambda}$$

where N_e is the species effective size, N is the breeding population size of a local population, n is the number of local populations, and λ is the per-generation probability of a population becoming extinct (and being replaced). The total breeding size of the species is $N_b = Nn$. This equation shows that the relationship between effective species size (N_e) and breeding species size (N_b) is rather complex. Our interest here is in the ratio of these two numbers, which given a little algebra, is expressed as

$$\frac{N_e}{N_b} = \frac{1-\lambda}{1 - F_{ST} + 2F_{ST}N\lambda}$$

45. Rogers and Jorde (1995) have suggested that our genetic estimates of 10,000 for species effective size are actually overestimates because, as shown by Nei and Takahata (1993), increased differentiation among populations (a larger F_{ST}) *inflates* the ratio of effective to breeding population size. Nei and Takahata suggest that the ratio is

$$\frac{N_e}{N_b} = \frac{1}{1 - F_{ST}}$$

This is true only in the special case in which there is no extinction, which can be seen by substituting $\lambda = 0$ into the Whitlock–Barton equation giving

$$\frac{N_e}{N_b} = \frac{1-0}{1 - F_{ST} + 2F_{ST}N0} = \frac{1}{1 - F_{ST}}$$

After λ increases to a certain level, the ratio of effective to breeding size will decline as either λ or F_{ST} increases. The critical value is given by solving for a ratio of 1, which occurs when

$$\lambda = \frac{F_{ST}}{1 + 2F_{ST}N}$$

For Fig. 7.12, the critical value is roughly $\lambda = 0.005$ for both F_{ST} values. Extinction rates higher than this increasingly reduce the ratio of effective to breeding population size.

46. Soltis et al. (1995).
47. Trinkaus (1995, p. 138).
48. Elise Eller, personal communication.

CHAPTER 8: NEANDERTAL DNA

1. Crichton (1990).
2. See Hagelberg (1994) and O'Rourke et al. (1996) for reviews of ancient DNA analysis.
3. The results of the first extraction of Neandertal DNA, from the Feldhofer specimen, are reported in Krings et al. (1997, 1999). The second extraction, from the Mezmaiskaya Cave specimen, is reported by Ovchinnikov et al. (2000).
4. Krings et al. (1997).
5. The date for the Feldhofer Neandertal specimen is from Larsen et al. (1998).
6. Krings et al. (1999).
7. Ovchinnikov et al. (2000).
8. If you look at Figure 2 in Ovchinnikov et al. (2000), you will be able to count 23 sequence differences between the Mesmaiskaya specimen and the human reference sequence, although the text indicates only 22 sequence differences. The number "22" is a typographic error, and should have read "23." (William Goodwin, personal communication, May 24, 2000).
9. Nordborg (1998).
10. Krings et al. (1997, p. 5583).
11. See Roberts et al. (1962) for the basic relationship between per-generation gene flow and accumulated ancestry over t generations. Also see Relethford (1999) for discussion of the application of this concept to the analysis of human evolution. The matrix of accumulated ancestry after t generations is

$$\mathbf{A} = \mathbf{M}^t$$

where $\mathbf{M}$ is the migration matrix and t is the number of generations.
12. Krings et al. (2000).
13. Adcock et al. (2001).
14. Relethford (2001a).

CHAPTER 9: PUTTING THE PIECES TOGETHER

1. Wolpoff (1999) among others already cited.
2. Stringer and Gamble (1993) among others already cited.
3. Some examples of biological distance analysis based on fossil samples include Stringer (1993, 1994), Waddle (1994), and Sokal et al. (1997).
4. Relethford (1999).
5. As discussed in Relethford (1999), the explanation for this effect is that, under a model of symmetric migrant *numbers*, the rows of the equilibrium matrix of accumulated ancestry will be equal to the relative weights of the populations (Jacquard 1974; Harpending and Ward 1982).
6. I am referring here specifically to regional continuity in neutral (nonadaptive) traits. The use of adaptive traits in analyses of population history introduces additional complica-

tions because distances between samples can also be affected by convergent (or divergent) changes due to natural selection. In my view, adaptive traits hold little information for this type of analysis, because selection can so easily mimic similarity due to shared ancestry.

7. Lahr (1994). See also Lahr (1996) for an expanded discussion.

8. See, for example, Wolpoff (1989, 1999), Wolpoff et al. (1984, 1994), Wolpoff and Caspari (1997), and Frayer et al. (1993, 1994) for summary reviews. Some specific case studies focusing on the Australasian evidence for continuity include Thorne and Wolpoff (1981), Kramer (1991), and Hawks, Oh et al. (2000). See Wolpoff et al. (2001) for recent evidence of continuity in both Australia and Europe.

9. See Frayer (1993, 1997) and Wolpoff (1999). A complete list of temporal changes in unique Neandertal traits and their persistence in post-Neandertal Europeans appears in Wolpoff (1999, p. 756).

10. Duarte et al. (1999).

11. Jorde et al. (1998, p. 134).

12. Thorne and Wolpoff (1981). See also Wolpoff et al. (1984) and Wolpoff and Caspari (1997).

13. Hawks, Hunley et al. (2000).

14. Smith (1994).

15. It is worth taking time to consider this population explosion in terms of the long-term *average* species effective size of 10,000 that has been repeatedly discussed. The figure of 10,000 is an average based on the time back to the coalescence of a given gene. Even if postexpansion effective size were quite large, the long-term *average* species effective size would still be relatively small (~10,000) if preexpansion effective size was small. As discussed in Chapter 7, the long-term average effective size is affected primarily by the smallest population size during an interval of time and is approximated by the harmonic mean size over time. To take an extreme case for illustrative purposes, imagine a species of effective size 5000 persisting for 2000 generations, at which point the species effective size increases to 500,000 and persists for another 2000 generations. The arithmetic mean over the past 4000 generations is over 250,000, but the harmonic mean is only 9900. The point here is that a species can have a history made up of period of small effective size followed by a period of large effective size, and the long-term average will lie much closer to the minimal value. The difference in conceptualizing species effective size lies in whether we are considering the effective size at any given point in time or the average over a long time.

16. See Relethford (1991a) for a general discussion of how changes in population size can affect F_{ST}. Although that paper was specifically geared to the study of short-term changes in human microevolution, the basic findings also apply here—given a certain amount of population growth, the continued effect of genetic drift will reduce and gene flow will lead to a reduction in F_{ST}. Over long enough intervals, gene flow can reduce differentiation among populations provided that sufficient population growth has taken place.

17. There is a long-standing argument in anthropology regarding the question of whether humans should be considered unique or considered as just another species. The dominant view has shifted back and forth over time. Studies of primate behavior have shown apes capable of some linguistic acquisition and possessing tool use and culture in the wild, behaviors that formerly had been considered uniquely human. Today, there is a tendency to consider humans as just another species in terms of evolutionary dynamics. In some ways, we may have gone too far down this particular road, because the past accomplishments of *Homo*, at least in terms of technology, are unparalleled in the rest of the world. I suggest that humans (in the broad sense) have always been mobile, inquisitive creatures that would not remain isolated long enough for speciation to have occurred

(although there remains the possibility that the Neandertals were an exception). I have no way of demonstrating to everyone's satisfaction that humans did not speciate during the past two million years, but I suggest the possibility that human evolutionary dynamics *might* have been different, and we need to consider the role of culture again.

18. Although I have discussed this model in terms of "sudden expansion" models, it could also fit the exponential growth model of Hawks, Hunley et al. (2000), which also predicts a rapid and accelerating growth of the human species.

19. Lahr and Foley (1994).

Bibliography

Adcock GJ, Dennis ES, Easteal S, Huntley GA, Jermin LS, Peacock WJ and Thorne A (2001) Mitochondrial DNA sequences in ancient Australians: Implications for modern human origins. *Proceedings of the National Academy of Sciences, USA* 98:537–542.

Aeillo LC (1993) The fossil evidence for modern human origins in Africa: A revised view. *American Anthropologist* 95:73–96.

Aitken MJ, Stringer CB and Mellars PA, eds. (1993) *The Origin of Modern Humans and the Impact of Chronometric Dating.* Princeton: Princeton University Press.

Ambrose SH (1998) Late Pleistocene human population bottlenecks, volcanic winter, and differentiation of modern humans. *Journal of Human Evolution* 34:623–651.

Awadalla P, Eyre-Walker A and Smith JM (1999) Linkage disequilibrium and recombination in hominid mitochondrial DNA. *Science* 286:2524–2525.

Awadalla P, Eyre-Walker A and Smith JM (2000) Questioning evidence for recombination in human mitochondrial DNA. *Science* 288:1931a (technical comment available online at www.sciencemag.org/cgi/content/full/288/5473/1931a).

Ayala FJ (1995) The myth of Eve: Molecular biology and human origins. *Science* 270:1930–1936.

Balter M and Gibbons A (2000) A glimpse of humans' first journey out of Africa. *Science* 288:948–950.

Bar-Yosef O (1994) The contributions of southwest Asia to the study of the origins of modern humans. In *Origins of Anatomically Modern Humans*, ed. by MH Nitecki and DV Nitecki, pp. 23–66. New York: Plenum Press.

Barbujani G, Magagni A, Minch E and Cavalli-Sforza LL (1997) An apportionment of human DNA diversity. *Proceedings of the National Academy of Sciences, USA* 94:4516–4519.

Barton NH and Whitlock MC (1997) The evolution of metapopulations. In *Metapopulation Biology: Ecology, Genetics, and Evolution*, ed. by IA Hanski and ME Gilpin, pp. 183–210. San Diego: Academic Press.

Batzer MA, Stoneking M, Alegria-Hartman M, Bazan H, Kass DH, Shaikh TH, Novick GE, Ioannou PA, Scheer WD, Herrera RJ and Deininger PL (1994) African origin of human-specific polymorphic *Alu* insertions. *Proceedings of the National Academy of Sciences, USA* 91:12288–12292.

Birdsell JB (1968) Some predictions for the Pleistocene based on equilibrium systems among recent hunter-gatherers. In *Man the Hunter*, ed. by RB Lee and I DeVore, pp. 229–240. Chicago: Aldine Publishing Company.

Bodmer WF and Cavalli-Sforza LL (1968) A migration matrix model for the study of random genetic drift. *Genetics* 59:565–592.

Bowcock AM, Kidd JR, Mountain JL, Hebert JM, Carotenuto L, Kidd KK and Cavalli-Sforza LL (1991) Drift, admixture, and selection in human evolution: A study with DNA polymorphisms. *Proceedings of the National Academy of Sciences, USA* 88:839–843.

Bowcock AM, Ruiz-Linares A, Tomfohrde J, Minch E, Kidd JR and Cavalli-Sforza LL (1994) High resolution of human evolutionary trees with polymorphic microsatellites. *Nature* 368:455–457.

Boyd R and Silk J (1999) *How Humans Evolved*. Second edition. New York: W.W. Norton & Company.

Bräuer G (1984) A craniological approach to the origin of anatomically modern *Homo sapiens* in Africa and implications for the appearance of modern humans. In *The Origins of Modern Humans: A World Survey of the Fossil Evidence*, ed. by FH Smith and F Spencer, pp. 327–410. New York: Alan R. Liss.

Bräuer G (1992) Africa's place in the evolution of *Homo sapiens*. In *Continuity or Replacement: Controversies in* Homo sapiens *Evolution*, ed. by G Bräuer and FH Smith, pp. 83–98. Rotterdam: A.A. Balkema.

Bräuer G and Smith FH (1992) *Continuity or Replacement: Controversies in* Homo sapiens *Evolution*. Rotterdam: A.A. Balkema.

Bräuer G and Stringer C (1997) Models, polarization, and perspectives on modern human origins. In *Conceptual Issues in Modern Human Origins Research*, ed. by GA Clark and CM Willermet, pp. 191–201. New York: Aldine de Gruyter.

Brown MH (1990) *The Search for Eve*. New York: Harper and Row Publishers.

Brown TA (1999) *Genomes*. New York: John Wiley & Sons.

Campbell BG and Loy JD (2000) *Humankind Emerging*. Eighth Edition. Boston: Allyn and Bacon.

Cann RL, Stoneking M and Wilson A (1987) Mitochondrial DNA and human evolution. *Nature* 325:31–36.

Cavalli-Sforza LL, Menozzi P and Piazza A (1994) *The History and Geography of Human Genes*. Princeton: Princeton University Press.

Clark GA (1997) Through a glass darkly: Conceptual issues in modern human origins research. In *Conceptual Issues in Modern Human Origins Research*, ed. by GA Clark and CM Willermet, pp. 60–76. New York: Aldine de Gruyter.

Clark GA and Willermet CM, eds. (1997) *Conceptual Issues in Modern Human Origins Research*. New York: Aldine de Gruyter.

Conroy GC (1997) *Reconstructing Human Origins: A Modern Synthesis*. New York: W.W. Norton & Company.

Crichton M (1990) *Jurassic Park*. New York: Alfred A. Knopf.

Crow JF and Kimura M (1970) *An Introduction to Population Genetics Theory*. Minneapolis: Burgess Publishing Company.

Deka R, Shriver MD, Yu LM, DeCroo S, Hundrieser J, Bunker CH, Ferrell RE and Chakraborty R (1995) Population genetics of dinucleotide $(dC - dA)_n \cdot (dG - dT)_n$ polymorphisms in world populations. *American Journal of Human Genetics* 56:461–474.

Devor EJ (1992) Introduction: A brief history of the RFLP. In *Molecular Applications in Biological Anthropology*, ed. by EJ Devor, pp. 1–18. Cambridge: Cambridge University Press.

Di Rienzo A, Donnelly P, Toomajian C, Sisk B, Hill A, Petzl-Erler ML, Haines GK and Barch DH (1998) Heterogeneity of microsatellite mutations within and between loci, and implications for human demographic histories. *Genetics* 148:1269–1284.

Donnelly P (1996) Interpreting genetic variability: The effects of shared evolutionary history. In *Variation in the Human Genome*, ed. by D Chadwick and G Cardew, pp. 25–50. Chichester, UK: John Wiley & Sons.

Donnelly P and Tavaré S (1995) Coalescents and genealogical structure under neutrality. *Annual Review of Genetics* 29:401–421.

Dorit RL, Akashi H and Gilbert W (1995) Absence of polymorphism at the ZFY locus on the human Y chromosome. *Science* 268:1183–1185.

Duarte C, Maurício J, Pettitt PB, Souto P, Trinkaus E, van der Plicht H and Zilhão J (1999) The early Upper Paleolithic human skeleton from the Abrigo do Lagar Velho (Portugal) and modern human emergence in Iberia. *Proceedings of the National Academy of Sciences, USA* 96:7604–7609.

Eller E (1999) Population substructure and isolation by distance in three continental regions. *American Journal of Physical Anthropology* 108:147–159.

Ereshefsky M, ed. (1992) *The Units of Evolution: Essays on the Nature of Species*. Cambridge, MA: MIT Press.

Excoffier L and Langaney A (1989) Origin of differentiation of human mitochondrial DNA. *American Journal of Human Genetics* 44:73–85.

Excoffier L and Schneider S (1999) Why hunter-gatherer populations do not show signs of Pleistocene demographic expansions. *Proceedings of the National Academy of Sciences, USA* 96:10597–10602.

Feder KL (2000) *The Past in Perspective: An Introduction to Human Prehistory*. Mountain View, CA: Mayfield Publishing Company.

Felsenstein J (1982) How can we infer geography and history from gene frequencies? *Journal of Theoretical Biology* 96:9–20.

Fischman J (1996) Evidence mounts for our African origins—and alternatives. *Science* 271:1364.

Frayer DW (1993) Evolution at the European edge: Neanderthal and Upper Paleolithic relationships. *Préhistore Européene* 2:9–69.

Frayer DW (1997) Perspectives on Neanderthals as ancestors. In *Conceptual Issues in Modern Human Origins Research*, ed. by GA Clark and CM Willermet, pp. 220–234. New York: Aldine de Gruyter.

Frayer DW, Wolpoff MH, Thorne AG, Smith FH and Pope GG (1993) Theories of modern human origins: The paleontological test. *American Anthropologist* 95:14–50.

Frayer DW, Wolpoff MH, Thorne AG, Smith FH and Pope GG (1994) Getting it straight. *American Anthropologist* 96:424–438.

Futuyma DJ (1997) *Evolutionary Biology*. Third edition. Sunderland, MA: Sinauer Associates.

Gabunia L, Vekua A, Lordkipanidze D, Swisher CC, Ferring R, Justus A, Nioradze M, Tvalchrelidze M, Antón SC, Bosinski G, Jöris O, de Lumley MA, Majsuradze G and Mouskhelishvili A (2000) Earliest Pleistocene hominid cranial remains from Dmanisi, Republic of Georgia: Taxonomy, geological setting, and age. *Science* 288:1019–1025.

Gagneux P, Wills C, Gerloff U, Tautz D, Morin PA, Boesch C, Fruth B, Hohmann G, Ryder OA and Woodruff DS (1999) Mitochondrial sequences show diverse evolutionary histories of African hominoids. *Proceedings of the National Academy of Sciences, USA* 96:5077–5082.

Gower JC (1966) Some distance properties of latent root and vector methods used in multivariate analysis. *Biometrika* 53:325–338.

Grün R and Thorne A (1997) Dating the Ngandong humans. *Science* 276:1575.

Hagelberg E (1994) Ancient DNA studies. *Evolutionary Anthropology* 2:199–207.

Hammer MF (1995) A recent common ancestry for human Y chromosomes. *Nature* 378:376–378

Hammer MF, Karafet T, Rasanayagam A, Wood ET, Altheide TK, Jenkins T, Griffiths RC, Templeton AR and Zegura SL (1998) Out of Africa and back again: Nested cladistic analysis of human Y chromosome variation. *Molecular Biology and Evolution* 15:427–441.

Hammer MF and Zegura SL (1996) The role of the Y chromosome in human evolutionary studies. *Evolutionary Anthropology* 5:116–134.

Hanski I (1998) Metapopulation dynamics. *Nature* 396:41–49.

Hanski IA and Giplin ME, eds. (1997) *Metapopulation Biology: Ecology, Genetics, and Evolution*. San Diego: Academic Press.

Harding RM (1996) Using the coalescent to interpret gene trees. In *Molecular Biology and Human Diversity*, ed. by AJ Boyce and CGN Mascie-Taylor, pp. 63–80. Cambridge: Cambridge University Press.

Harding RM (1997) Lines of descent from mitochondrial Eve: An evolutionary look at coalescence. In *Progress in Population Genetics and Human Evolution*, ed. by P. Donnelly and S. Tavaré, pp. 15–31. New York: Springer-Verlag.

Harding RM, Fullerton SM, Griffiths RC, Bond J, Cox MJ, Schneider JA, Moulin DS and Clegg JB (1997) Archaic African *and* Asian lineages in the genetic ancestry of modern humans. *American Journal of Human Genetics* 60:772–789.

Harpending H (1974) Genetic structure of small populations. *Annual Review of Anthropology* 3:229–243.

Harpending HC (1994) Signature of ancient population growth in a low-resolution mitochondrial DNA mismatch distribution. *Human Biology* 66:591–600.

Harpending HC, Batzer MA, Gurven M, Jorde LB, Rogers AR and Sherry ST (1998) Genetic traces of ancient demography. *Proceedings of the National Academy of Sciences, USA* 95:1961–1967.

Harpending H and Jenkins (1973) Genetic distance among Southern African populations. In *Methods and Theories of Anthropological Genetics*, ed. by MH Crawford and PL Workman, pp. 177–199. Albuquerque: University of New Mexico Press.

Harpending H, Relethford JH and Sherry ST (1996) Methods and models for understanding human diversity. In *Molecular Biology and Human Diversity*, ed. by AJ Boyce and CGN Mascie-Taylor, pp. 283–299. Cambridge: Cambridge University Press.

Harpending HC, Sherry ST, Rogers AR and Stoneking M (1993) The genetic structure of ancient human populations. *Current Anthropology* 34:483–496.

Harpending HC and Ward R (1982) Chemical systematics and human populations. In *Biochemical Aspects of Evolutionary Biology*, ed. by M Nitecki, pp. 213–256. Chicago: University of Chicago Press.

Harris EE and Hey J (1999) X chromosome evidence for ancient human histories. *Proceeding of the National Academy of Sciences, USA* 96:3320–3324.

Hartl DL and Clark AG (1997) *Principles of Population Genetics*. Third edition. Sunderland, MA: Sinauer Associates.

Hassan FA (1981) *Demographic Archaeology*. New York: Academic Press.

Hawks J, Hunley K, Lee, S-H and Wolpoff M (2000) Population bottlenecks and Pleistocene human evolution. *Molecular Biology and Evolution* 17:2–22.

Hawks J, Oh S, Hunley K, Dobson S, Cabana G, Dayalu P and Wolpoff MH (2000) An Australasian test of the recent African origin theory using the WLH-50 calvarium. *Journal of Human Evolution* 39:1–22.

Hedges SB, Jumar S, Tamura K and Stoneking M (1992) Human origins and analysis of mitochondrial DNA sequences. *Science* 255:737–739.

Hedrick PW and Gilpin ME (1997) Genetic effective size of a metapopulation. In *Metapopulation Biology: Ecology, Genetics, and Evolution*, ed. by IA Hanski and ME Gilpin, pp. 165–182. San Diego: Academic Press.

Hey J (1997) Mitochondrial and nuclear genes present conflicting portraits of human origins. *Molecular Biology and Evolution* 14:166–172.

Hooton EA, Dupertuis CW and Dawson H (1955) *The Physical Anthropology of Ireland*. Papers of the Peabody Museum, Vol. 30, Nos. 1–2. Cambridge, MA: Peabody Museum.

Horai S, Hayasaka K, Tsugane K and Takahata N (1995) Recent African origin of modern humans revealed by complete sequences of hominoid mitochondrial DNA. *Proceedings of the National Academy of Sciences, USA* 92:532–536.

Hou Y, Potts R, Yuan B, Guo Z, Deino A, Wang W, Clark J, Xie G and Huang W (2000) Mid-pleistocene Acheulian-like stone technology of the Bose Basin, South China. *Science* 287:1622–1625.

Howell FC (1996) Thoughts on the study and interpretation of the human fossil record. In *Contemporary Issues in Human Evolution*, ed. by WE Meikle, FC Howell and NG Jablonski, pp. 1–46. California Academy of Sciences Memoir 21. San Francisco: California Academy of Sciences.

Howells WW (1976) Explaining modern man: Evolutionists *versus* migrationists. *Journal of Human Evolution* 5:477–495.

Huang W, Fu Y-X, Chang BHJ, Gu X, Jorde LB and Li W-H (1998) Sequence variation in ZFX introns in human populations. *Molecular Biology and Evolution* 15:138–142.

Hudson RR (1990) Gene genealogies and the coalescent process. In *Oxford Surveys in Evolutionary Biology, Volume 7*, ed. by D. Futuyma and J. Antonovics, pp. 1–44. Oxford: Oxford University Press.

Imaizumu Y, Morton NE and Harris DE (1970) Isolation by distance in artificial populations. *Genetics* 66:569–582.

Imaizumi Y, Morton NE and Lalouel JM (1973) Kinship and race. In *Genetic Structure of Populations*, ed. by NE Morton, pp. 228–233. Honolulu: University Press of Hawaii.

Ingman M, Kaessmann H, Pääblo S and Gyllensten U (2000) Mitochondrial genome variation and the origin of modern humans. *Nature* 408:708–713.

Jacquard A (1974) *The Genetic Structure of Populations*. New York: Springer-Verlag.

Jorde LB (1980) The genetic structure of subdivided human populations: A review. In *Current Developments in Anthropological Genetics, Volume 1: Theory and Methods*, ed. by JH Mielke and MH Crawford, pp. 135–208. New York: Plenum Press.

Jorde LB and Bamshad M (2000) Questioning evidence for recombination in human mitochondrial DNA. *Science* 288:1931a (technical comment available online at www.sciencemag.org/cgi/content/full/288/5473/1931a).

Jorde LB, Bamshad M and Rogers AR (1998) Using mitochondrial and nuclear DNA markers to reconstruct human evolution. *BioEssays* 20:126–136.

Jorde LB, Bamshad MJ, Watkins WS, Zenger R, Fraley AE, Krakowiak PA, Carpenter KD, Soodyall H, Jenkins R and Rogers AR (1995) Origins and affinities of modern humans: A comparison of mitochondrial and nuclear genetic data. *American Journal of Human Genetics* 57:523–538.

Jorde LB, Rogers AR, Bamshad M, Watkins WS, Krakowiak P, Sung S, Kere J and Harpending HC (1997) Microsatellite diversity and the demographic history of modern humans. *Proceedings of the National Academy of Sciences, USA* 94:3100–3103.

Jurmain R, Nelson H, Kilgore L and Trevathan W (1999) *Introduction to Physical Anthropology*. Eighth edition. Belmont, CA: Wadsworth Publishing Company.

Kaessmann H, Wiebe V and Pääbo S (1999) Extensive nuclear DNA sequence diversity among chimpanzees. *Science* 286:1159–1162.

Kimmel M, Chakraborty R, King JP, Bamshad M, Watkins WS and Jorde LB (1998) Signatures of population expansion in microsatellite repeat data. *Genetics* 148:1921–1930.

King W (1864) The reputed fossil man of the Neanderthal. *Quarterly Journal of Science* 1:88–97.

Kivisild T and Villems R (2000) Questioning evidence for recombination in human mitochondrial DNA. *Science* 288:1931a (technical comment available online at www.sciencemag.org/cgi/content/full/288/5473/1931a).

Klein RG (1999) *The Human Career: Human Biological and Cultural Origins*. Second edition. Chicago: University of Chicago Press.

Kramer A (1991) Modern human origins in Australasia: Replacement or evolution? *American Journal of Physical Anthropology* 86:455–473.

Kramer A (1993) Human taxonomic diversity in the Pleistocene: Does *Homo erectus* represent multiple hominid species? *American Journal of Physical Anthropology* 91:161–171.

Krings M, Capelli C, Tschentscher F, Geisert H, Meyer S, von Haeseler A, Grossschmidt K, Possnert G, Paunovic M and Pääbo S (2000) A view of Neandertal genetic diversity. *Nature Genetics* 26:144–146.

Krings M, Geisert H, Schmitz RW, Krainitzki H and Pääbo S (1999) DNA sequence of the mitochondrial hypervariable region II from the Neandertal type specimen. *Proceedings of the National Academy of Sciences, USA* 96:5581–5585.

Krings M, Stone A, Schmitz RW, Krainitzki H, Stoneking M and Pääbo S (1997) Neandertal DNA sequences and the origin of modern humans. *Cell* 90:19–30.

Kumar S, Hedrick P, Dowling T and Stoneking M (2000) Questioning evidence for recombination in human mitochondrial DNA. *Science* 288:1931a (technical comment available online at www.sciencemag.org/cgi/content/full/288/5473/1931a).

Lahr MM (1994) The multiregional model of modern human origins: A reassessment of the morphological basis. *Journal of Human Evolution* 26:23–56.

Lahr MM (1996) *The Evolution of Modern Human Diversity: A Study of Cranial Variation*. Cambridge: Cambridge University Press.

Lahr MM and Foley R (1994) Multiple dispersals and modern human origins. *Evolutionary Anthropology* 3:48–60.

Larson CS, Matter RM and Gebo DL (1998) *Human Origins: The Fossil Record*. Third edition. Prospect Heights, IL: Waveland Press.

Latter BDH (1980) Genetic differences within and between populations of the major human groups. *American Naturalist* 116:220–237.

Lewin R (1997) *Bones of Contention: Controversies in the Search for Human Origins*. Second edition. Chicago: University of Chicago Press.

Lewontin RC (1972) The apportionment of human diversity. *Evolutionary Biology* 6:381–398.

Li W-H (1977) Distribution of nucleotide differences between two randomly chosen cistrons in a finite population. *Genetics* 85:331–337.

Livi-Bacci M (1997) *A Concise History of World Population*. Second edition. Malden, MA: Blackwell Publishers.

Livshits G and Nei M (1990) Relationships between intrapopulational and interpopulational genetic diversity in man. *Annals of Human Biology* 6:501–513.

Loewe L and Scherer S (1997) Mitochondrial Eve: The plot thickens. *Trends in Ecology and Evolution* 12:422–423.

Lynch M (1989) Phylogenetic hypotheses under the assumption of neutral quantitative-genetic variation. *Evolution* 43:1–17.

Maddison DR (1991) African origin of human mitochondrial DNA reexamined. *Systematic Zoology* 40:355–363.

Marjoram P and Donnelly P (1997) Human demography and the time since mitochondrial Eve. In *Progress in Population Genetics and Human Evolution*, ed. by P. Donnelly and S. Tavaré, pp. 107–131. New York: Springer-Verlag.

Maryuma T and Kimura M (1980) Genetic variability and effective population size when local extinction and recolonization of subpopulations is frequent. *Proceedings of the National Academy of Sciences, USA* 77:6710–6714.

Meikle WE and Parker ST (1994) *Naming Our Ancestors: An Anthology of Hominid Taxonomy*. Prospect Heights, IL: Waveland Press.

Mellars PA and Stringer CB, eds. (1989) *The Human Revolution: Behavioral and Biological Perspectives on the Origin of Modern Humans*. Princeton: Princeton University Press.

Morton NE (1973) Prediction of kinship from a migration matrix. In *Genetic Structure of Populations*, ed. by NE Morton, pp. 119–123. Honolulu: University Press of Hawaii.

Nei M (1978) The theory of genetic distance and the evolution of human races. *Japanese Journal of Human Genetics* 23:341–369.

Nei M (1987) *Molecular Evolutionary Genetics*. New York: Columbia University Press.

Nei M and Grauer D (1984) Extent of protein polymorphism and the neutral mutation theory. *Evolutionary Biology* 27:73–118.

Nei M and Livshits G (1989) Genetic relationships of Europeans, Asians and Africans and the origin of modern *Homo sapiens*. *Human Heredity* 39:276–281.

Nei M and Roychoudhury AK (1982) Genetic relationship and evolution of human races. *Evolutionary Biology* 14:1–59.

Nei M and Takahata N (1993) Effective population size, genetic diversity, and coalescence time in subdivided populations. *Journal of Molecular Evolution* 37:240–244.

Nitecki MH and Nitecki DV, eds. (1994) *Origins of Anatomically Modern Humans*. New York: Plenum Press.

Nordborg M (1998) On the probability of Neanderthal ancestry. *American Journal of Human Genetics* 63:1237–1240.

North KE, Martin LJ and Crawford MH (2000) The origin of the Irish Travellers and the genetic structure of Ireland. *Annals of Human Biology* 27:453–465.

O'Brien SJ, Wildt DE, Goldman D, Merril CR and Bush M (1983) The cheetah is depauperate in genetic variation. *Science* 221:459–462.

O'Rourke DH, Carlyle SW and Parr RL (1996) Ancient DNA: Methods, progress, and perspectives. *American Journal of Human Biology* 8:557–571.

Omoto K and Tobias PV, eds. (1998) *The Origins and Past of Modern Humans—Towards Reconciliation*. Singapore: World Scientific Publishing Company.

Ovchinnikov IV, Götherström A, Romanova GP, Kharitonov VM, Lidén K and Goodwin W (2000) Molecular analysis of Neanderthal DNA from the northern Caucasus. *Nature* 404:490–493.

Park MA (1998) *Biological Anthropology*. Second edition. Mountain View, CA: Mayfield Publishing Company.

Parsons TJ and Irwin JA (2000) Questioning evidence for recombination in human mitochondrial DNA. *Science* 288:1931a (technical comment available online at www.sciencemag.org/cgi/content/full/288/5473/1931a).

Penny D, Steel M, Waddell PJ and Hendy MD (1995) Improved analyses of human mtDNA sequences support a recent African origin for *Homo sapiens*. *Molecular Biology and Evolution* 12:863–882.

Pérez-Lezaun A, Calafell F, Mateu E, Comas D, Ruiz-Pacheco R and Bertranpetit J (1997) Microsatellite variation and the differentiation of modern humans. *Human Genetics* 99:1–7.

Poirier FE and McKee JK (1999) *Understanding Human Evolution*. Fourth edition. Upper Saddle River, NJ: Prentice Hall.

Rampino MR and Self S (1992) Volcanic winter and accelerated glaciation following the Toba super-eruption. *Nature* 359:50–52.

Reich DE and Goldstein DB (1998) Genetic evidence for a Paleolithic human population expansion in Africa. *Proceedings of the National Academy of Sciences, USA* 95:8119–8123.

Relethford JH (1991a) Effects of changes in population size on genetic microdifferentiation. *Human Biology* 63:629–641.

Relethford JH (1991b) Genetic drift and anthropometric variation in Ireland. *Human Biology* 63:155–165.

Relethford JH (1994) Craniometric variation among modern human populations. *American Journal of Physical Anthropology* 95:53–62.

Relethford JH (1995) Genetics and modern human origins. *Evolutionary Anthropology* 4:53–63.

Relethford JH (1997a) Hemispheric difference in human skin color. *American Journal of Physical Anthropology* 104:449–457.

Relethford JH (1997b) Mutation rate and excess African heterozygosity. *Human Biology* 69:785–792.

Relethford JH (1998a) Genetics of modern human origins and diversity. *Annual Review of Anthropology* 27:1–23.

Relethford JH (1998b) Mitochondrial DNA and ancient population growth. *American Journal of Physical Anthropology* 105:1–7.

Relethford JH (1999) Models, predictions, and the fossil record of modern human origins. *Evolutionary Anthropology* 8:7–10.

Relethford JH (2000a) Human skin color diversity is highest in sub-Saharan African populations. *Human Biology* 72:773–780.

Relethford JH (2000b) *The Human Species: An Introduction to Biological Anthropology*. Fourth edition. Mountain View, CA: Mayfield Publishing Company.

Relethford JH (2001a) Ancient DNA and the origin of modern humans. *Proceedings of the National Academy of Sciences, USA* 98:390–391.

Relethford JH (2001b) Genetic history of the human species. In *Handbook of Statistical Genetics*, ed. by D Balding, M Bishop and C Cannings. Chichester, UK: John Wiley & Sons (in press).

Relethford JH and Blangero J (1990) Detection of differential gene flow from patterns of quantitative variation. *Human Biology* 62:5–25.

Relethford JH and Crawford MH (1995) Anthropometric variation and the population history of Ireland. *American Journal of Physical Anthropology* 96:25–38.

Relethford JH, Crawford MH and Blangero J (1997) Genetic drift and gene flow in post-famine Ireland. *Human Biology* 69:443–465.

Relethford JH and Harpending HC (1994) Craniometric variation, genetic theory, and modern human origins. *American Journal of Physical Anthropology* 95:249–270.

Relethford JH and Harpending HC (1995) Ancient differences in population size can mimic a recent African origin of modern humans. *Current Anthropology* 36:667–674.

Relethford JH and Jorde LB (1999) Genetic evidence for larger African population size during recent human evolution. *American Journal of Physical Anthropology* 108:251–260.

Rightmire GP (1990) *The Evolution of* Homo erectus: *Comparative Anatomical Studies of an Extinct Human Species*. Cambridge: Cambridge University Press.

Rightmire GP (1992) *Homo erectus*: Ancestor or evolutionary side branch? *Evolutionary Anthropology* 1:43–49.

Roberts DF and Hiorns RW (1962) The dynamics of racial intermixture. *American Journal of Human Genetics* 14:261–277.

Rogers AR (1992) Error introduced by the infinite-sites model. *Molecular Biology and Evolution* 9:1181–1184.

Rogers AR (1995) Genetic evidence for a Pleistocene population explosion. *Evolution* 49:608–615.

Rogers AR (1997) Population structure and modern human origins. In *Progress in Population Genetics and Human Evolution*, ed. by P Donnelly and S Tavaré, pp. 55–79. New York: Springer-Verlag.

Rogers AR, Fraley AE, Bamshad MJ, Watkins WS and Jorde LB (1996) Mitochondrial mismatch analysis is insensitive to the mutational process. *Molecular Biology and Evolution* 13:895–902.

Rogers AR and Harpending HC (1986) Migration and genetic drift in human populations. *Evolution* 40:1312–1327.

Rogers AR and Harpending H (1992) Population growth makes waves in the distribution of pairwise genetic differences. *Molecular Biology and Evolution* 9:552–569.

Rogers AR and Jorde LB (1995) Genetic evidence on modern human origins. *Human Biology* 67:1–36.

Rogers AR and Jorde LB (1996) Ascertainment bias in estimates of average heterozygosity. *American Journal of Human Genetics* 58:1033–1041.

Romesburg HC (1984) *Cluster Analysis for Researchers*. Belmont, CA: Wadsworth Publishing.

Roychoudhury AK and Nei M (1988) *Human Polymorphic Genes: World Distribution*. New York: Oxford University Press.

Ruff CB, Trinkaus E and Holliday TW (1997) Body mass and encephalization in Pleistocene *Homo*. *Nature* 387:173–176.

Russell, PJ (1996) *Genetics*. Fourth edition. New York: HarperCollins College Publishers.

Ruvolo M (1996) A new approach to studying modern human origins. *Molecular Phylogenetics and Evolution* 5:202–219.

Ruvolo M, Zehr S, von Dornum M, Pan D, Chang B and Lin J (1993) Mitochondrial COII sequences and modern human origins. *Molecular Biology and Evolution* 10:1115–1135.

Ryman N, Chakraborty R and Nei M (1983) Differences in the relative distribution of human gene diversity between electrophoretic red and white cell antigen loci. *Human Heredity* 33:93–102.

Sagan C (1977) *The Dragons of Eden: Speculations on the Evolution of Human Intelligence*. New York: Ballantine Books.

Sarich VM (1971) A molecular approach to the question of human origins. In *Back-ground for Man: Readings in Physical Anthropology*, ed. by P Dolhinow and V Sarich, pp. 60–81. Boston: Little, Brown and Company.

Sarich VM and Wilson A (1967a) Rates of albumin evolution in primates. *Proceedings of the National Academy of Sciences, USA* 58:142–148.

Sarich VM and Wilson A (1967b) Immunological time scale for hominoid evolution. *Science* 158:1200–1203.

Schick KD and Toth N (1993) *Making Silent Tools Speak: Human Evolution and the Dawn of Technology*. New York: Simon and Schuster.

Schopf JW, ed. (1992) *Major Events in the History of Life*. Boston: Jones and Bartlett Publishers.

Sherry ST and Batzer MA (1997) Modeling human evolution—To tree or not to tree? *Genome Research* 7:947–949.

Sherry ST, Harpending HC, Batzer MA and Stoneking M (1997) *Alu* evolution in human populations: Using the coalescent to estimate effective population size. *Genetics* 147:1977–1982.

Sherry ST, Rogers AR, Harpending H, Soodyall H, Jenkins T and Stoneking M (1994) Mismatch distributions of mtDNA reveal recent human population expansions. *Human Biology* 66:761–775.

Shreeve J (1995) *The Neandertal Enigma: Solving the Mystery of Modern Human Origins*. New York: William Morrow and Company.

Simmons T (1994) Archaic and modern *Homo sapiens* in the contact zones: Evolutionary schematics and model predictions. In *Origins of Anatomically Modern Humans*, ed. by MH Nitecki and DV Nitecki DV, pp. 201–225. New York: Plenum Press.

Slatkin M (1977) Gene flow and genetic drift in a species subject to frequent local extinction. *Theoretical Population Biology* 12:253–262.

Smith CAB (1969) Local fluctuations in gene frequencies. *Annals of Human Genetics* 32:251–260.

Smith FH (1994) Samples, species, and speculations in the study of modern human origins. In *The Origin of Anatomically Modern Humans*, ed. by MH Nitecki and DV Nitecki, pp. 227–249. New York: Plenum Press.

Smith FH, Falsetti AB and Donnelly SM (1989) Modern human origins. *Yearbook of Physical Anthropology* 32:35–68.

Smith FH and Spencer F, eds. (1984) *The Origins of Modern Humans: A World Survey of the Fossil Evidence*. New York: Alan R. Liss.

Smith FH, Trinkaus E, Pettitt PB, Karavanić I and Paunović (1999) Direct radiocarbon dates for Vindija G_1 and Velika Pećina Late Pleistocene hominid remains. *Proceedings of the National Academy of Sciences, USA* 96:12281–12286.

Smith SL and Harrold FB (1997) A paradigm's worth of difference? Understanding the impasse over modern human origins. *Yearbook of Physical Anthropology* 40:113–138.

Sokal RR, Oden NL, Walker J and Waddle DM (1997) Using distance matrices to choose between competing theories and an application to the origin of modern humans. *Journal of Human Evolution* 32:501–522.

Soltis J, Boyd R and Richardson PJ (1995) Can group-functional behaviors evolve by cultural group selection? An empirical test. *Current Anthropology* 36:473–494.

Stein PL and Rowe BM (1999) *Physical Anthropology*. Seventh edition. New York: McGraw-Hill.

Stoneking M (1993) DNA and recent human evolution. *Evolutionary Anthropology* 2:60–73.

Stoneking M (1994) In defense of "Eve"—A response to Templeton's critique. *American Anthropologist* 96:131–141.

Stoneking M and Cann RL (1989) African origin of human mitochondrial DNA. In *The Human Revolution: Behavioral and Biological Perspectives on the Origin of Modern Humans*, ed. by PA Mellars and CB Stringer, pp. 17–30. Princeton: Princeton University Press.

Stoneking M, Fontius JJ, Clifford SL, Soodyall H, Arcot SS, Saha N, Jenkins T, Tahir MA, Deininger PL and Batzer MA (1997) *Alu* insertion polymorphisms and human evolution: Evidence for a larger population size in Africa. *Genome Research* 7:1061–1071.

Stoneking M, Sherry ST, Redd AJ and Vigilant L (1992) New approaches to dating suggest a recent age for the human mtDNA ancestor. *Philosophical Transactions of the Royal Society of London B* 337:167–175.

Strachan T and Read AP (1999) *Human Molecular Genetics*. Second edition. New York: John Wiley & Sons.

Stringer CB (1985) Middle Pleistocene hominid variability and the origin of Late Pleistocene humans. In *Ancestors: The Hard Evidence*, ed. by E Delson, pp. 289–295. New York: Alan R. Liss.

Stringer CB (1990) The emergence of modern humans. *Scientific American* 263(6):98–104.

Stringer CB (1993) Reconstructing human evolution. In *The Origin of Modern Humans and the Impact of Chronometric Dating*, ed. by MJ Aitken, CB Stringer and PA Mellars, pp. 179–195. Princeton: Princeton University Press.

Stringer CB (1994) Out of Africa—A personal history. In *Origins of Anatomically Modern Humans*, ed. by MH Nitecki and DV Nitecki, pp. 149–172. New York: Plenum Press.

Stringer CB and Andrews P (1988) Genetic and fossil evidence for the origin of modern humans. *Science* 239:1263–1268.

Stringer C and Bräuer G (1994) Methods, misreading, and bias. *American Anthropologist* 96:416–424.

Stringer C and Gamble C (1993) *In Search of the Neanderthals: Solving the Puzzle of Human Origins*. New York: Thames and Hudson.

Stringer C and McKie R (1996) *African Exodus: The Origins of Modern Humanity*. New York: Henry Holt and Company.

Swisher CC, Curtis GH, Jacob T, Getty AG, Suprijo A and Widiasmoro (1994) Age of the earliest known hominids in Java, Indonesia. *Science* 263:1118–1121.

Swisher CC, Rink WJ, Antón SC, Schwartz HP, Curtis GH, Suprijo A and Widiasmoro (1996) Latest *Homo erectus* of Java: Potential contemporaneity with *Homo sapiens* in southeast Asia. *Science* 274:1870–1874.

Swisher CC, Rink WJ, Schwartz HP and Antón S (1997) Dating the Ngandong humans. *Science* 276:1575–1576.

Takahata N (1993) Allelic geneology and human evolution. *Molecular Biology and Evolution* 10:2–22.

Takahata N (1994) Repeated failures that led to the eventual success in human evolution. *Molecular Biology and Evolution* 11:803–805.

Takahata N (1995) A genetic perspective on the origin and history of humans. *Annual Review of Ecology and Systematics* 26:343–372.

Tattersall I (1986) Species recognition in human paleontology. *Journal of Human Evolution* 15:165–175.

Tattersall I (1992) Species concepts and species identification in human evolution. *Journal of Human Evolution* 22:341–349.

Tattersall I (1994a) How does evolution work? *Evolutionary Anthropology* 3:2–3.

Tattersall I (1994b) Morphology and phylogeny. *Evolutionary Anthropology* 3:40–41.

Tattersall I (1995) *The Fossil Trail: How We Know What We Think We Know About Human Evolution*. New York: Oxford University Press.

Templeton AR (1992) Human origins and analysis of mitochondrial DNA sequences. *Science* 255:737.

Templeton AR (1993) The "Eve" hypotheses: A genetic critique and reanalysis. *American Anthropologist* 95:51–72.

Templeton AR (1994) "Eve": Hypothesis compatibility versus hypothesis testing. *American Anthropologist* 96:141–147.

Templeton AR (1996) Contingency tests of neutrality using intra/interspecific gene trees: The rejection of neutrality for the evolution of the mitochondrial cytochrome oxidase II gene in the hominoid primates. *Genetics* 144:1263–1270.

Templeton AR (1997a) Out of Africa? What do the genes tell us? *Current Opinion in Genetics & Development* 7:841–847.

Templeton AR (1997b) Testing the Out of Africa replacement hypothesis with mitochondrial DNA data. In *Conceptual Issues in Modern Human Origins Research*, ed. by GA Clark and CM Willermet, pp. 329–360.

Templeton AR (1998) Human races: A genetic and evolutionary perspective. *American Anthropologist* 100:632–650.

Thomson R, Pritchard JK, Shen P, Oefner PJ and Feldman MW (2000) Recent common ancestry of human Y chromosomes: Evidence from DNA sequence data. *Proceedings of the National Academy of Sciences, USA* 97:7360–7365.

Thorne AG and Wolpoff MH (1981) Regional continuity in Australasian Pleistocene hominid evolution. *American Journal of Physical Anthropology* 55:337–349.

Thorne AG and Wolpoff MH (1992) The multiregional evolution of humans. *Scientific American* 266(4):76–83.

Thorne AG, Wolpoff MH and Eckhardt RB (1993) Genetic variation in Africa. *Science* 261:1507–1508.

Tierney J, Wright L and Springen (1988) The search for Adam and Eve. *Newsweek* January 11, 1998:46–52.

Tishkoff SA, Dietzsch E, Speed W, Pakstis AJ, Kidd JR, Cheung K, Bonné-Tamir B, Santachiara-Benerecetti AS, Moral P, Krings M, Pääbo S, Watson E, Risch N, Jenkins T and Kidd KK (1996) Global patterns of linkage disequilibrium at the CD4 locus and modern human origins. *Science* 271:1380–1387.

Trinkaus E (1995) Neanderthal mortality patterns. *Journal of Archaeological Science* 22:121–142.

Trinkaus E and Shipman P (1992) *The Neandertals: Changing the Image of Mankind*. New York: Alfred A. Knopf.

Underhill PA, Jin L, Lin AA, Mehdi SQ, Jenkins T, Vollrath D, Davis RW, Cavalli-Sforza LL and Oefner PJ (1997) Detection of numerous Y chromosome biallelic polymor-phisms by denaturing high-performance liquid chromatography. *Genome Research* 7:996–1005.

Vigilant L, Stoneking M, Harpending H, Hawkes K and Wilson AC (1991) African popula-tions and the evolution of human mitochondrial DNA. *Science* 253:1503–1507.

Waddle DM (1994) Matrix correlation tests support a single origin for modern humans. *Nature* 368:452–454.

Wainscoat J (1987) Out of the Garden of Eden. *Nature* 325:13.

Walker A and Shipman P (1996) *The Wisdom of the Bones: In Search of Human Origins*. New York: Alfred A. Knopf.

Weiss KM (1984) On the numbers of members of the genus *Homo* who have ever lived, and some evolutionary implications. *Human Biology* 56:637–649.

Weiss KM (1988) In search of times past: Gene flow and invasion in the generation of human diversity. In *Biological Aspects of Human Migration*, ed. by CGN Mascie-Taylor and GW Lasker, pp. 130–166. Cambridge: Cambridge University Press.

Weiss KM and Maryuma T (1976) Archeology, population genetics, and studies of human racial ancestry. *American Journal of Physical Anthropology* 44:31–50.

Whitlock MC and Barton NH (1997) The effective size of a subdivided population. *Genetics* 146:427–441.

Whittam TS, Clark AG, Stoneking M, Cann RL and Wilson AC (1986) Allelic variation in human mitochondrial genes based on patterns of restriction site polymorphisms. *Proceedings of the National Academy of Sciences, USA* 83:9611–9615.

Willermet CM and Hill B (1997) Fuzzy set theory and its implications for speciation models. In *Conceptual Issues in Modern Human Origins Research*, ed. by GA Clark and CM Willermet, pp. 77–88. New York: Aldine de Gruyter.

Williams RC (1989) Restriction fragment length polymorphisms (RFLP). *Yearbook of Physical Anthropology* 32:159–184.

Williams-Blangero S and Blangero J (1989) Anthropometric variation and the genetic structure of the Jirels of Nepal. *Human Biology* 61:1–12.

Wills C (1995) When did Eve live? An evolutionary detective story. *Evolution* 49:593–607.

Wise CA, Sraml M, Rubinsztein DC and Easteal S (1997) Comparative nuclear and mitochondrial genome diversity in humans and chimpanzees. *Molecular Biology and Evolution* 14:707–716.

Wobst HM (1974) Boundary conditions for Paleolithic social systems: A simulation approach. *American Antiquity* 39:147–178.

Wolpoff MH (1989) Multiregional evolution: The fossil alternative to Eden. In *The Human Revolution: Behavioral and Biological Perspectives on the Origin of Modern Humans*, ed. by PA Mellars and CB Stringer, pp. 62–108. Princeton: Princeton University Press.

Wolpoff MH (1994a) How does evolution work? *Evolutionary Anthropology* 3:4–5.

Wolpoff MH (1994b) Time and phylogeny. *Evolutionary Anthropology* 3:38–39.

Wolpoff MH (1999) *Paleoanthropology*. Second edition. Boston: McGraw-Hill.

Wolpoff MH and Caspari R (1997) *Race and Human Evolution*. New York: Simon & Schuster.

Wolpoff MH, Hawks J and Caspari R (2000) Multiregional, not multiple origins. *American Journal of Physical Anthropology* 112:129–136.

Wolpoff, MH, Hawks J, Frayer DW and Hunley K (2001) Modern human ancestry at the peripheries: A test of the replacement theory. *Science* 291:293–297.

Wolpoff MH, Thorne AG, Jelínek J and Yinyun Z (1993) The case for sinking *Homo erectus*: 100 years of *Pithecanthropus* is enough! In *100 Years of* Pithecanthropus*: The* Homo erectus *Problem*, ed. by JL Franzen. *Courier Forschunsinstitut Senckenberg* 17:341–361.

Wolpoff MH, Wu X and Thorne AG (1984) Modern *Homo sapiens* origins: A general theory of hominid evolution involving the fossil evidence from East Asia. In *The Origins of Modern Humans: A World Survey of the Fossil Evidence*, ed. by FH Smith and F Spencer, pp. 411–483. New York: Alan R. Liss.

Wolpoff MH, Thorne AG, Smith FH, Frayer DW and Pope GG (1994) Multiregional evolution: A world-wide source for modern human populations. In *The Origin of Anatomically Modern Humans*, ed. by MH Nitecki and DV Nitecki, pp. 175–199. New York: Plenum Press.

Wood B and Collard M (1999) The changing face of genus *Homo*. *Evolutionary Anthropology* 8:195–207.

Workman PL, Harpending H, Lalouel JM, Lynch C, Niswander JD and Singleton R (1973) Population studies on southwestern Indian tribes. VI. Papago population structure: A comparison of genetic and migration analyses. In *Genetic Structure of Populations*, ed. by NE Morton, pp. 166–194. Honolulu: University Press of Hawaii.

Wright S (1940) Breeding structure of populations in relation to speciation. *American Naturalist* 74:232–248.

Wright S (1951) The genetical structure of populations. *Annals of Eugenics* 15:323–354.

Wright S and Provine WB, ed. (1986) *Evolution: Selected Papers*. Chicago: University of Chicago Press.

Yellen JE, Brooks AS, Cornelissen E, Mehlman MJ and Stewart K (1995) A Middle Stone Age worked bone industry from Katanda, Upper Semliki Valley, Zaire. *Science* 268:553–556.

Zubrow E (1989) The demographic modelling of Neanderthal extinction. In *The Human Revolution: Behavioral and Biological Perspectives on the Origin of Modern Humans*, ed. by PA Mellars and CB Stringer, pp. 212–231. Princeton: Princeton University Press.

Index